Functions of Two Variables

Second Edition

CHAPMAN & HALL/CRC MATHEMATICS

OTHER CHAPMAN & HALL/CRC MATHEMATICS TEXTS:

Functions of Two Variables,
Second edition
S. Dineen

Network Optimization
V. K. Balakrishnan

Sets, Functions and Logic:
A foundation course in mathematics
Second edition
K. Devlin

Algebraic Numbers and Algebraic
Functions
P. M. Cohn

Dynamical Systems:
Differential equations, maps and
chaotic behaviour
D. K. Arrowsmith and C. M. Place

Control and Optimization
B. D. Craven

Elements of Linear Algebra
P. M. Cohn

Error-Correcting Codes
D. J. Bayliss

Introduction to Calculus of
Variations
U-Brechtken-Mandershneid

Intergration Theory
W. Filter and K. Weber

Algebraic Combinatories
C. D. Godsil

An Introduction to Abstract
Analysis-PB
W. A. Light

The Dynamic Cosmos
M. Madsen

Algorithms for Approximation II
J. C. Mason and M. G. Cox

Introduction to Combinatorics
A. Slomson

Galois Theory
I. N. Stewart

Elements of Algebraic Coding
Theory
L. R. Vermani

Linear Algebra:
A geometric approach
E. Sernesi

A Concise Introduction to
Pure Mathematics
M. W. Liebeck

Geometry of Curves
J. W. Rutter

Full information on the complete range of Chapman & Hall/CRC Mathematics books is available from the publishers

Functions of Two Variables

Second Edition

Seán Dineen

CHAPMAN & HALL/CRC

Boca Raton London New York Washington, D.C.

Library of Congress Cataloging-in-Publication Data

Dineen, Seán, 1944-
    Functions of two variables / Seán Dineen—2nd ed.
       p. cm. (Chapman & Hall/CRC Mathematics)
    Includes index.
    ISBN 1-58488-190-9 (alk. paper)
    1. Calculus. I. Title. II. Series.
QA303 .D628 2000
515′.94—dc21   00-043021

This book contains information obtained from authentic and highly regarded sources. Reprinted material is quoted with permission, and sources are indicated. A wide variety of references are listed. Reasonable efforts have been made to publish reliable data and information, but the authors and the publisher cannot assume responsibility for the validity of all materials or for the consequences of their use.

Neither this book nor any part may be reproduced or transmitted in any form or by any means, electronic or mechanical, including photocopying, microfilming, and recording, or by any information storage or retrieval system, without prior permission in writing from the publisher.

The consent of CRC Press LLC does not extend to copying for general distribution, for promotion, for creating new works, or for resale. Specific permission must be obtained in writing from CRC Press LLC for such copying.

Direct all inquiries to CRC Press LLC, 2000 N.W. Corporate Blvd., Boca Raton, Florida 33431.

Trademark Notice: Product or corporate names may be trademarks or registered trademarks, and are used only for identification and explanation, without intent to infringe.

Visit the CRC Press Web site at www.crcpress.com

© 2000 by Chapman & Hall/CRC

No claim to original U.S. Government works
International Standard Book Number 1-58488-190-9
Library of Congress Card Number 00-043021
Printed in the United States of America   4 5 6 7 8 9 0
Printed on acid-free paper

To SARA

who likes mathematics and little books

Contents

Preface	ix
1 Functions from R^2 to R	1
2 Partial Derivatives	8
3 Critical Points	14
4 Maxima and Minima	21
5 Saddle Points	29
6 Sufficiently Regular Functions	37
7 Linear Approximation	45
8 Tangent Lines	54
9 Method and Examples of Lagrange Multipliers	62
10 Theory and Examples of Lagrange Multipliers	72
11 Tangent Planes	81
12 The Chain Rule	87
13 Directed Curves	98
14 Curvature	106
15 Quadratic Approximation	118
16 Vector Valued Differentiation	125
17 Complex Analysis	132
18 Line Integrals	138
19 The Fundamental Theorem of Calculus	148
20 Double Integrals	156
21 Coordinate Systems	165
22 Green's Theorem	175
Solutions	185
Index	188

Preface

This book was initially based on a short course of 20 lectures, given to second year students at University College Dublin during the autumn of 1992. Later, two chapters on integration theory were added to improve the balance between differential and integral calculus. The students had completed a one-year course on differential and integral calculus for real valued functions of one real variable—this is the prerequisite for reading this book—and this course was designed as an introduction to the calculus of several variables.

My initial motivation for writing this book was to provide my own students with a friendly set of notes that they could read in their entirety. As the book took shape, I realized that I was addressing the subject in a manner somewhat different from the standard texts on several variable calculus. It is difficult to explain precisely why this occurred. Nevertheless, an attempted explanation may also help you, the reader, in your approach and I will try to give a partial one.

Research mathematicians typically spend their working lives doing research, learning new mathematics and teaching. They teach themselves new mathematics mainly to further their own research. Yet, often their own way of learning mathematics is the complete opposite of the way they present mathematics to their students. On approaching a new area of mathematics, the research mathematician is usually looking for some result (or technique). He or she will generally not know precisely what is being sought and lives in hope that by searching, often backwards and forwards through a text, the required result will somehow be recognized. The search through the literature will neither be random nor logical, but will be based on accumulated experience and intuition. Once the objective has been identified the research mathematician works backwards to satisfy professional standards for a precise meaning of the terms involved and the context in which the result may be applied. Finally, and this depends on many things, the research mathematician may even decide to satisfy a need for certainty and will then work through the background proofs. Thus the mathematician, when doing research, behaves like a detective and in fact there is no alternative since the plot is not revealed until the story is almost over. Nevertheless, with students we first reveal the climax (theorem), then the evidence (proof) and finally the intuition (explanation and examples). This robs the subject of its excitement and does not use the students' own intuition and experience. I have tried to approach the material of these lectures as a research mathematician approaches research: full of doubt, more intuitively than logically, somewhat imprecise about where we may be going, but with a general objective in mind, moving backwards and forwards, trying simple cases, using various tricks that have previously proved useful, experimenting and eventually arriving at something of interest. Having obtained useful results intuitively, I have returned to justify them mathematically. At this stage the reasoning behind the proofs is often more acceptable and the proofs themselves become an integral part of a unified process by adding to our understanding of the applications, by showing the usefulness of earlier theoretical results and by suggesting further developments. Of course, I have not fully succeeded in this attempt, but feel nevertheless that I have gone some way in this direction. I believe that this is almost the only way to learn mathematics and that most students are trying to follow this approach.

Although the calculus of several variables is often presented as a fully mature subject in its own right, it is clear that most of the concepts are the natural evolution of attempting to imitate the one-dimensional theory and I have tried to follow this approach in my presentation. The restriction to functions of two variables simplifies the notation and at the same time introduces most of the main concepts that arise in higher dimensions. I believe that a clear understanding of the two-variables case is a suitable introduction to the higher dimensional situation. I have tried to be both rigorous and self-contained and so have clearly marked out assumptions made and discussed the significance of results used without proof.

We discuss all possible functions which involve two variables and so look at functions from $\mathbf{R}^2 \to \mathbf{R}$, $\mathbf{R} \to \mathbf{R}^2$ and $\mathbf{R}^2 \to \mathbf{R}^2$. This provides a basic introduction to three subjects, i.e., calculus of several variables, differential geometry and complex analysis.

In the first 12 chapters we discuss maxima and minima of functions of two variables on both open sets and level curves. Second order derivatives and the Hessian are used to find extremal values on open sets, while the method of Lagrange multipliers is developed for level curves. In the process we introduce partial derivatives, directional derivatives, the gradient, critical points, tangent planes, the normal line and the chain rule and also discuss regularity conditions such as continuity of functions and their partial derivatives and the relationship between differentiation and approximation. In Chapters 13 to 16 we investigate the curvature of plane curves. Chapters 18 to 22 are devoted to integration theory on $\mathbf{R}^2$. We study Fubini's theorem (on the change of order of integration), line and area integrals and connect them using Green's theorem. In Chapter 17 we introduce holomorphic (or **C**-differentiable) functions and using approximation methods derive the Cauchy-Riemann equations. This introduction to complex analysis is augmented in the final chapter where Green's theorem is combined with the Cauchy-Riemann equations to prove Cauchy's theorem. Partial derivatives enter into and play an important role in every topic discussed.

As life is not simple, many things depend on more than one variable and it is thus not surprising that the methods developed in this book are widely used in the physical sciences, economics, statistics and engineering. We mention some of these applications and give a number of examples in the text.

Anyone interested in several variable calculus will profit from reading this book. Students suddenly exposed to the multidimensional situation in its full generality will find a gentle introduction here. Students of engineering, economics and science who ask simple but fundamental questions will find some answers and, perhaps, even more questions here.

This book may be used fully or partially as the basis for a course in which a lecturer has the option of inserting extra material and developing more fully certain topics. Alternatively it can be used as supplementary reading for courses on advanced calculus or for self study. The material covered in each chapter can be presented in approximately 60 minutes, although in some of my lectures I was not able to cover fully all examples in the allocated time, and for aesthetical and mathematical reasons I have, in writing up these lectures, sometimes ignored the time frame imposed by the classroom.

It is a real pleasure to acknowledge the help I have received in bringing this project to fruition. Siobhán Purcell displayed patience, understanding and skill in preparing the written text. Alun Carr, Peter O'Neill and Brendan Quigley gave valuable advice on preparing the diagrams, while William Aliaga-Kelly, Derek O'Connor, Michael Mackey and Richard Timoney performed the onerous task of producing the diagrams. Maciej Klimek my former colleague (now at Uppsala), Milne Anderson (London) and Pauline Mellon (Dublin) provided valuable advice and sustained support at all stages. The mathematics editor at Chapman & Hall, Achi Dosanjh, gave encouragement and practical support at crucial times. To all these and to my students at University College Dublin, I offer sincere thanks. I hope you, the reader, will enjoy this book and I welcome your comments. I can be contacted at the address below. The second edition contains corrections, some small additions and some rearrangement of the first edition.

Seán Dineen, Department of Mathematics, University College Dublin, Belfield, Dublin 4 (Ireland)

1
Functions from $\mathbf{R}^2$ to $\mathbf{R}$

Summary. *We introduce the problem of finding the maximum and minimum of a real-valued function of two variables. The one-dimensional theory suggests that we discuss the problem for functions defined on open sets and their boundaries. We define open sets, consider a possible definition of the derivative and define the graph of a function of two variables.*

In this course we will discuss all possible functions which involve two variables and so look at functions from $\mathbf{R}^2$ into $\mathbf{R}$, from $\mathbf{R}$ into $\mathbf{R}^2$ and from $\mathbf{R}^2$ into $\mathbf{R}^2$. We begin by considering functions from $\mathbf{R}^2$ into $\mathbf{R}$ and our objective is to obtain methods for finding maxima and minima. If the functions are arbitrarily behaved we get nowhere, so we have to make some assumptions—we will use the general term **regularity conditions**—on the functions considered. These regularity conditions usually relate to continuity and differentiability. First, however, we try and see, based on our one-dimensional experience, how we might proceed and then return to look more closely at what we need in order to proceed. The main one-dimensional motivation is the following fundamental result.

Theorem 1. *If $f:[a,b] \longrightarrow \mathbf{R}$ is continuous on the closed interval $[a,b]$, then f has a maximum and a minimum on $[a,b]$.*

In other words there are two points in $[a,b]$, x_1 and x_2, such that
$$f(x_1) \leq f(x) \leq f(x_2)$$
for all x in $[a,b]$.

The function f has a minimum value $f(x_1)$ which is achieved at x_1 and a maximum value $f(x_2)$ which is achieved at x_2 (Figure 1.1). Both the maximum and minimum are finite. The maximum (or minimum) may appear in two ways:

(i) it may occur at a point inside $[a, b]$, i.e., in (a, b) or
(ii) it may occur at a boundary point, i.e., at either a or b.

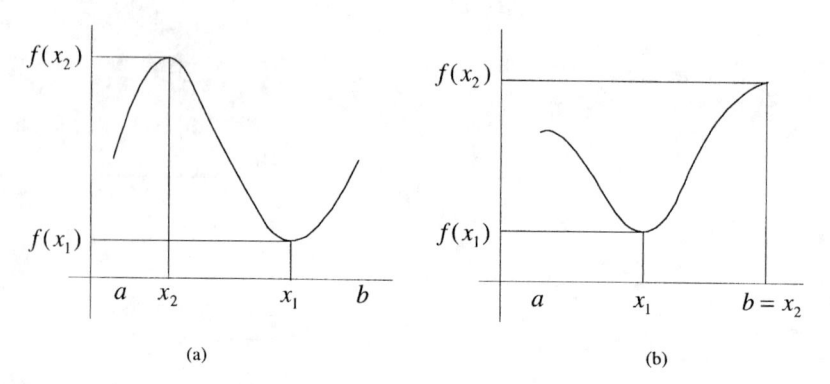

Figure 1.1

In Figure 1.1(a) we see that possibility (i) occurs for both maximum and minimum, while in (b) possibility (ii) occurs for the maximum and possibility (i) occurs for the minimum.

If f is differentiable on (a, b) and the maximum occurs inside, then we have $f'(x_2) = 0$ so our method of proceeding in this case is to look at all x in (a, b) where $f'(x) = 0$. We call these the **critical points** of f. There is usually only a small number of critical points so we can evaluate f at these points. This takes care of all possibilities inside and since there are only two other points—the end points a and b—we can find $f(a)$ and $f(b)$ and locate the maximum by choosing which one of this small set of possibilities gives the largest value of the function.

If we are now considering a function f defined on some subset U of $\mathbf{R}^2$, it is natural to attempt to try and break the problem (of finding the maximum) into two problems. To deal with f "inside" we have to define what the derivative of f might mean.

From the one-dimensional theory we see that $f'(x)$ only makes sense if f is defined at all points near x and, indeed, any method of deciding that a point x is a maximum or minimum of the function f must in some way involve the values of f at all points near x. So we first need to define a suitable analogue of an **open interval** in $\mathbf{R}^2$. We are thus led to the following definition.

Definition 2. *A subset U of $\mathbf{R}^2$ is open if for each point (x_0, y_0) in U there exists $\delta > 0$ such that*

$$\{(x, y); (x - x_0)^2 + (y - y_0)^2 < \delta^2\} \subset U.$$

So a set is open if each point in the set can be **surrounded** by a **small disc** of **positive** radius which still lies in the set.

Functions from $\mathbf{R}^2$ to $\mathbf{R}$

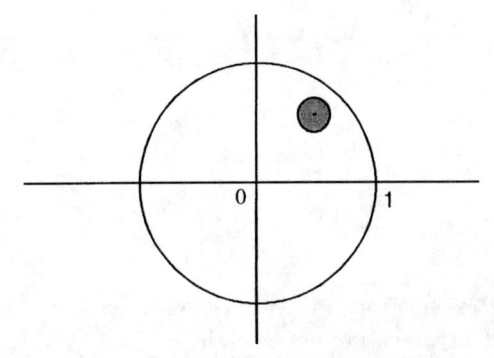

Figure 1.2

Example 3. $\{(x,y); x^2 + y^2 < 1\}$ is open (Figure 1.2).

Example 4. $\{(x,y); x^2 + y^2 = 1\}$ is not open (Figure 1.3).

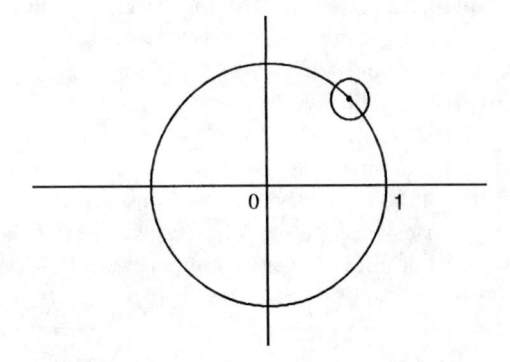

Figure 1.3

Example 5. $\{(x,y); x^2 + y^2 \leq 1\}$ is not open (Figure 1.4). Points P inside are all right but points Q on the boundary are not.

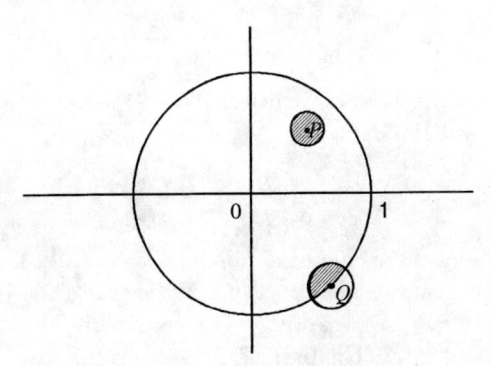

Figure 1.4

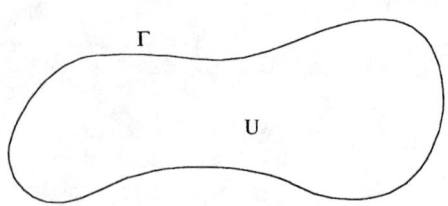

Figure 1.5

Notice that the open set in the above examples has $<$ signs in its definition while the sets which are not open have $\leq$ or $=$. Why?

We also see immediately a new complication in finding the maximum if it occurs on the boundary.

The boundary of $[a,b]$ consisted of just two points, whereas the boundary of $U \subset \mathbf{R}^2$ consists of a curve Γ (see Figure 1.5)—which contains an infinite number of points—and so the method of evaluating the function at each boundary point is impossible—there are too many points—and a new method has to be devised. We do this later—it is called the method of **Lagrange multipliers**.

So far we have only defined a set on which we might hope to differentiate. Now we will try to define the derivative of f and afterwards try to see what we might mean by a critical point. The task of a critical point is to help locate maxima and minima. From the one variable theory we have

$$f'(x_0) = \lim_{x \to x_0} \frac{f(x) - f(x_0)}{x - x_0} = \lim_{\Delta x \to 0} \frac{f(x_0 + \Delta x) - f(x_0)}{\Delta x}.$$

We consider the following possibility:

$$\lim_{(x,y) \to (x_0,y_0)} \frac{f(x,y) - f(x_0,y_0)}{(x,y) - (x_0,y_0)}$$

$$= \lim_{(\Delta x, \Delta y) \to (0,0)} \frac{f(x_0 + \Delta x, y_0 + \Delta y) - f(x_0, y_0)}{(\Delta x, \Delta y)}.$$

We immediately run into difficulties. Apart altogether from the possible definitions of limit the expression

$$\frac{f(x,y) - f(x_0,y_0)}{(x,y) - (x_0,y_0)}$$

does not make sense. We cannot divide by a vector. We could identify (x,y) with the complex number $z = x + iy$ and we may then divide by complex numbers. This in fact is interesting when considering functions into $\mathbf{R}^2$ and we will look into it later (Chapter 17). However, for real-valued functions it would lead to a definition in which the only differentiable functions are the

Functions from $\mathbf{R}^2$ to $\mathbf{R}$

constant functions (see Example 62) and this is clearly of no use in finding maxima and minima. Why? Moreover, if we move up one dimension to $\mathbf{R}^3$ how do we divide by $(\Delta x, \Delta y, \Delta z)$? This won't work. We return to fundamentals and try another approach.

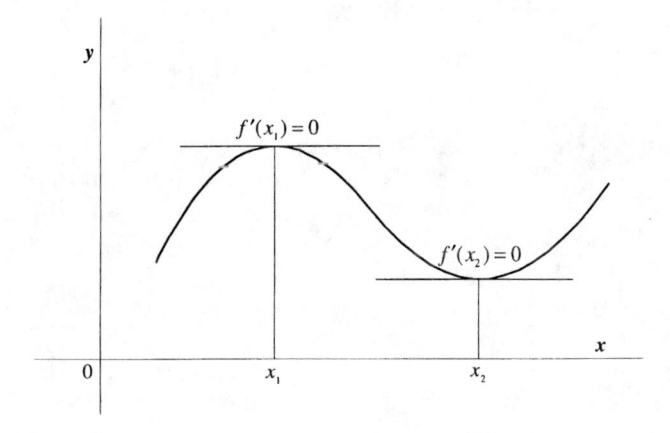

Figure 1.6

In the one-dimensional theory we are also led to critical points by considering the **graph** and noting that the line that fits closest to the graph—the **tangent line**—is horizontal at all critical points (Figure 1.6) and that in general the slope of this line is the derivative (Figure 1.7).

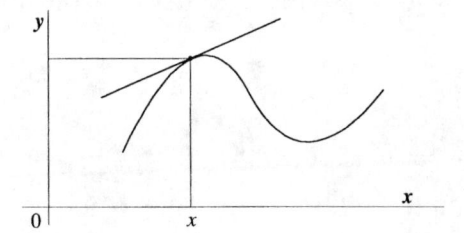

Figure 1.7

We will try to draw the graph to see if it leads to anything of interest.

The graph of the function of one variable $y = f(x)$ consists of all points in $\mathbf{R}^2$ of the form $(x, f(x))$ or (x, y) where $y = f(x)$.

Definition 6. *If $f : U \subset \mathbf{R}^2 \to \mathbf{R}$, then the graph of f consists of the set of all points*

$$\{(x, y, f(x, y)); (x, y) \subset U\}.$$

Since we are already using x and y to denote variables we often let $z = f(x, y)$ and then we are considering the points (x, y, z) where $z = f(x, y)$.

Exercises

1.1 Find the natural domain of definition of the following functions (i.e., find where the functions make sense):

(a) $f(x, y) = \dfrac{y + 2}{x}$
(b) $f(u, v) = \dfrac{uv}{u - 2v}$
(c) $f(r, s) = \big(\log(r/s)\big) - \sqrt{1 - r}$.

1.2 Sketch the function $f : \mathbf{R} \to \mathbf{R}$, $f(x) = x^3 - 3x$.
Find, by inspection, intervals $[a, b]$, $[c, d]$ and $[e, f]$ such that

(a) f achieves its maximum and minimum on $[a, b]$ at a and b, respectively,

(b) f achieves its maximum on $[c, d]$ both at the point d and at some point in (c, d),

(c) f achieves both its maximum and minimum on $[e, f]$ at precisely two points.

1.3 Let $f(x, y) = \dfrac{x^2}{4} + \dfrac{y^2}{9}$ and $U = \{(x, y); f(x, y) < 1\}$. Sketch the set U.
Show that the point $(-1, 2)$ lies in U. Find a positive δ such that
$$\{(x, y); x^2 + (y - 1)^2 < \delta^2\} \subset U.$$
Show that U is open. What formula describes the boundary of U?

1.4 Give an example of an open set in $\mathbf{R}^2$ whose boundary consists of two points.

1.5 Show that the function
$$f(x) = 2x^3 - 9x^2 + 12x$$
has a local maximum and a local minimum, but no maximum and no minimum on the set $(0, 3)$.

1.6 By using one variable calculus find the minimum of the function
$$f(x) = x^2 - 10x + 9, \quad x \in \mathbf{R}.$$
Sketch the graph of f.

1.7 Let $f : \mathbf{R} \to \mathbf{R}$ denote a function with continuous first and second derivatives and suppose $f''(x) \neq 0$ for all $x \in \mathbf{R}$. By drawing sketches convince yourself that f has a minimum on $\mathbf{R}$ in each of the following cases:

Functions from $\mathbf{R}^2$ *to* $\mathbf{R}$

(a) f has one local minimum and no local maximum,
(b) f has precisely two local minima and one local maximum.

1.8 On which of the following sets has the function $f(x) = x^3$ (a) a maximum and (b) a minimum?
   (i) $\{x; 0 < x \leq 1\}$ (ii) $\{x; 0 \leq x \leq 1\}$
   (iii) $\{x; 0 < x < 1\}$ (iv) $\{x; 0 \leq x < 1\}$
   (v) $\{x; 0 < x < \infty\}$ (vi) $\{x; 0 \leq x < \infty\}$
   (vii) $\mathbf{R}$ (viii) $\mathbf{N}$ (the natural numbers $1, 2, 3, \ldots$)
   (ix) $\mathbf{Q}^+ \cup \{0\}$, i.e., the non-negative rational numbers.

1.9 Which of the following sets are open in $\mathbf{R}^2$?
   (a) $\{(x, y); 0 < x < 1, y > 0\}$,
   (b) $\{(x, y); x^2 + y^2 < 2x\}$,
   (b) $\{(x, y); xy$ is rational $\}$,
   Sketch the sets in (a) and (b) and find their boundaries.

1.10 If U is an open subset of $\mathbf{R}^2$ and A is a finite subset of U show that $U \backslash A$ is open.

1.11 If $f : (a, b) \to \mathbf{R}$ has continuous first and second derivatives show that f attains its maximum over (a, b) at $c \in (a, b)$ if $f'(c) = 0$ and $f''(x) \geq 0$ for all $x \in (a, b)$.

2
Partial Derivatives

Summary. *We define and examine level sets of the graph and arrive at the concept of the partial derivative. Examples of partial derivatives are given.*

The graph of $f: U \subset \mathbf{R}^2 \to \mathbf{R}$ is a subset of $\mathbf{R}^3$ and we will call it a **surface**. As we get to study graphs we will see that they have many of the features that we intuitively associate with surfaces and so our use of the word surface is not unreasonable. Since the graph is a subset of $\mathbf{R}^3$ a certain amount of ingenuity is required in order to obtain a faithful sketch and a number of standard approaches to this problem have been developed over the years. One method is to consider **cross sections** and by examining sufficiently many cross sections we may get ideas on where the maximum or minimum might be located.

A cross section of $\mathbf{R}^2$ is a **line** and this is determined by a linear equation in two variables $ax + by = c$. A cross section of $\mathbf{R}^3$ is a **plane** and this is determined by a linear equation in three variables

$$ax + by + cz = d$$

where a, b, c and d are real numbers, and by varying these numbers we get different cross sections. A cross section of a surface consists of the points on the surface which satisfy the linear equation.

There are many choices for a, b, c and d. We will follow a general principle in mathematics—take the easiest cases first and examine them. If we are lucky we get what we want. If we are not lucky we at least have some experience when we have to consider more complicated cases. A second principle also comes in here—put as many constants as possible equal to 0 in the first instance and after that take as many as possible equal to 1. I say "as many as possible" since if we are too enthusiastic we end up with a situation that is completely trivial.

Partial Derivatives 9

We start by looking at the cross section given by the equation $z = d$, i.e., we take $a = b = 0$ and $c = 1$ in the linear equation.

So our cross section of the graph consists of points of the form

$$f(x, y) = d.$$

If the function is sufficiently regular we get a curve called a **level set** or **level curve** of f.

Example 7.

Figure 2.1

This is precisely the type of diagram we come across in maps and weather charts and it is helpful to keep these examples in mind—in maps the **level sets** are called **contours**. On a map $f(x, y)$ is the height above sea level of the point (x, y). In the above picture of Benbulben (Figure 2.1), we have indicated by arrows two level sets created by nature. A third level set is the base of the mountain.

Example 8. $f(x, y) = (x^2 + y^2)^{1/2}$ (see Figure 2.2). Since $f(x, y) \geq 0$ for all x, y we only consider level sets of the form $(x^2 + y^2)^{1/2} = d$ where d is positive. This is easily seen to be a circle of radius d.

From a brief inspection and our own previous experience of map reading we are able to write down immediately a number of general rules regarding level sets. We summarize these as follows.

1. The level sets $f(x, y) = d_1$ and $f(x, y) = d_2$ cannot cross if $d_1 \neq d_2$ since if they did the function f would take two different values at the point of intersection. This is impossible. Why?
2. On a level set of a function the function takes precisely one value—this is just restating the definition of level set.

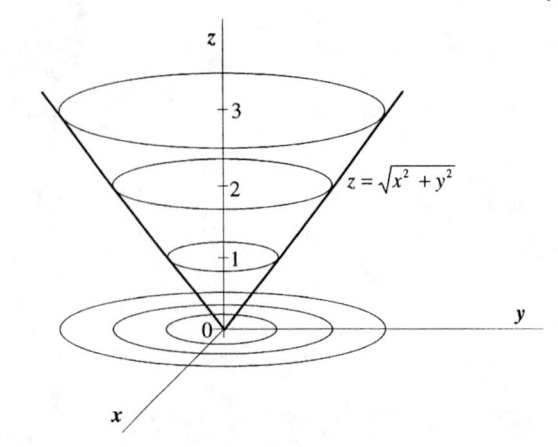

Figure 2.2

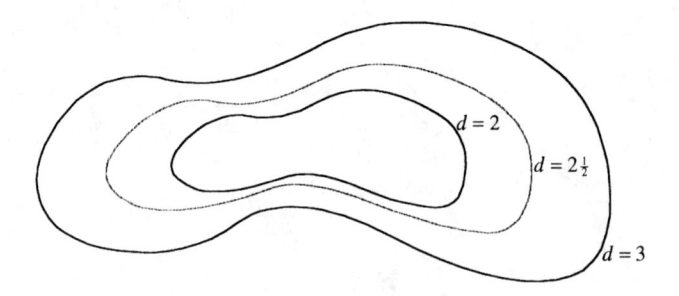

Figure 2.3

3. If our function is sufficiently regular (e.g., continuous) then between the level sets $f(x,y) = 2$ and $f(x,y) = 3$ we expect to find the level set $f(x,y) = 2\frac{1}{2}$ (Figure 2.3).

This is why we have to make assumptions about functions. Of course we could draw more and more level sets and get more and more information, but eventually the diagram would be too crowded. So we have a choice, either we examine every level set of the function or we make assumptions about the function. The fewer assumptions we make the larger the set of functions to which we can apply the theory. On the other hand, the more assumptions we make the stronger the conclusions we may derive. Good mathematics consists of choosing the right balance.

4. To locate the maximum we move from the level set $f(x,y) = 2$ to $f(x,y) = 3$ and then on to $f(x,y) = 4$ until we find that there is no level set taking a higher value.

In other words, we follow a path by crossing level sets with increasing constants (for the level sets) until we reach the maximum (Figure 2.4). If

Partial Derivatives

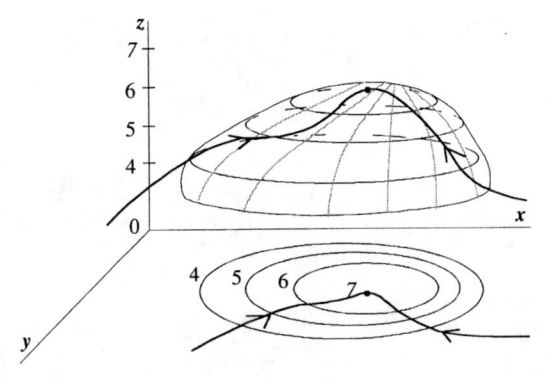

Figure 2.4

you follow this method for $f(x,y) = 4 - x^2 - y^2$ you will find easily that f has a maximum at the origin and no minimum. There are of course many paths leading to the maximum but from the mathematical point of view it is easiest to go, for instance, in an east–west direction (keep the map reading analogy in mind).

To become more mathematical let us suppose that the function $f(x,y)$ has a maximum at the point (a,b) (Figure 2.5).

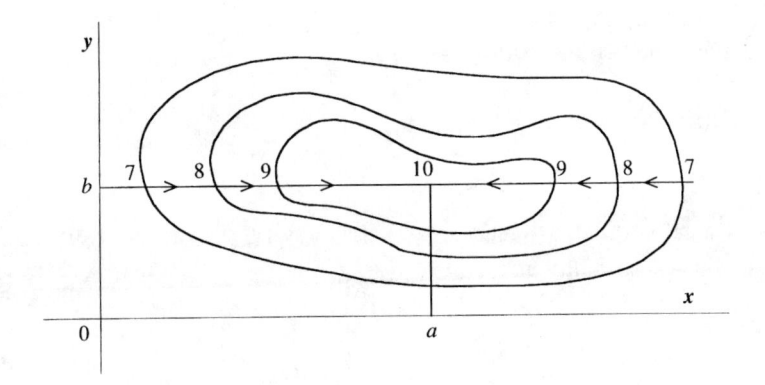

Figure 2.5

On the east–west path we are following the function which has the value $f(x,b)$ and we reach the maximum when $x = a$. Now $f(x,b)$ involves only one variable—x—so we should be able to consider it as a function of one variable. To emphasize this and in order to write it in the one variable language with which we are familiar, we define the function g as

$$g(x) = f(x,b).$$

Then, if f has a maximum at (a,b), the function g will have a maximum

at a. If g is twice differentiable, then the one variable theory tells us that
$$g'(a) = 0 \quad \text{and} \quad g''(a) \leq 0.$$
We have
$$g'(a) = \lim_{h \to 0} \frac{g(a+h) - g(a)}{h} = \lim_{h \to 0} \frac{f(a+h,b) - f(a,b)}{h}$$
and we are led to the following definition.

Definition 9.
$$\frac{\partial f}{\partial x}(a,b) = \lim_{h \to 0} \frac{f(a+h,b) - f(a,b)}{h}.$$

This is called the **partial derivative of f with respect to x at (a,b)** and is also written $f_x(a,b)$.

We use ∂ instead of d only to remind us that we are dealing with a function of **more** than one variable. If we compare this with our first attempt at differentiation we note that it is rather similar but we have changed the denominator to get a scalar and now it is possible to divide. In practice the computation of $\dfrac{\partial f}{\partial x}$ is very easy—you just treat the y variable as a constant. Since this derivative was obtained by differentiating a function of one variable, g, it is clear that the ordinary rules of differentiation for one variable—the product rule, etc.—apply.

Example 10. $f(x,y) = xy$.
$$\frac{\partial f}{\partial x} = y.$$

If you have difficulty, initially you might try the following. Take a specific constant say 12 (not 0 or 1 in this case) and put $y = 12$. Then $f(x, 12) = 12x$ and so
$$\frac{\partial f}{\partial x}(x, 12) = \frac{d}{dx}(12x) = 12.$$
Now replace 12 by y on both sides and we get $\dfrac{\partial f}{\partial x}(x,y) = y$.

Example 11. $f(x,y) = \sin(xy)$.
$$\frac{\partial f}{\partial x}(x,y) = y \cos(xy).$$

Exercises

2.1 Using scales where (i) 1 x-unit = 1 y-unit, (ii) 1 x-unit = 2 y-units,

Partial Derivatives

(iii) 1 y-unit $= 2$ x-units, sketch four level sets of each of the following functions:

  (a) $f(x,y) = x^2 + y^2 - 1$
  (b) $f(u,v) = 6 - 2u - 3v$
  (c) $f(x,y) = \sqrt{x^2 + 4y^2 + 25}$
  (d) $f(s,t) = 4s^2 + t^2$.

Comment on the relationships between the level sets in (a), (c), and (d).

2.2 Find $\dfrac{\partial f}{\partial x}$ for the following functions:

  (a) $f(x,y) = 2x^4 y^3 - xy^2 + 3y + 1$
  (b) $f(x,y) = \dfrac{x}{y} - \dfrac{y}{x}$
  (c) $f(x,y) = xe^y + y\sin x$.

2.3 Show that every level set of the function $f(x,y) = xy$ lies on a level set of the function $g(x,y) = \dfrac{2(xy-3)^2}{(xy)^4 + 1}$. Find the real number c such that the level set $xy = 1$ is contained in the level set $g(x,y) = c$.

2.4 Let $f(x,y) = x^2 + xy + y^2 - 4y$. Find the level sets of f which contain
(a) $(0,0)$ (b) $(1,1)$, (c) $(1,2)$ (d) $(-1,1)$.

3
Critical Points

Summary. *We define the gradient, and critical points are introduced as the zeros of the gradient. We show that all local maxima and minima are critical points. Directional derivatives are defined and their relationship with partial derivatives outlined. We use vectors and the scalar product to simplify our notation.*

So, if we approach the maximum in an east–west direction, we are led to the concept of the partial derivative $\dfrac{\partial f}{\partial x}$ and at the maximum point (a, b) we have $\dfrac{\partial f}{\partial x}(a, b) = 0$. Naturally, we can do exactly the same if we approach the maximum in a north–south direction (Figure 3.1).

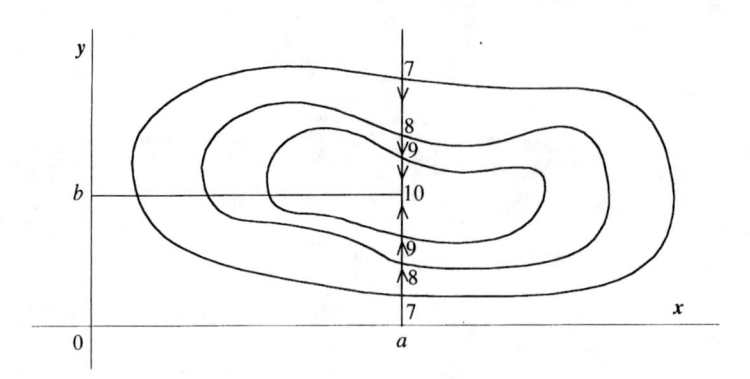

Figure 3.1

Critical Points

In this case the variable x is kept fixed and we allow y to vary. We define the function k by $k(y) = f(a, y)$. If f has a maximum at (a, b), then k has a maximum at b and if k is differentiable, then $k'(b) = 0$.

We have

$$k'(y) = \lim_{h \to 0} \frac{k(y+h) - k(y)}{h} = \lim_{h \to 0} \frac{f(a, y+h) - f(a, y)}{h}$$

and we are again led to a definition.

Definition 12. *If f is defined on an open subset of $\mathbf{R}^2$ containing (a, b), then*

$$\frac{\partial f}{\partial y}(a, b) = \lim_{h \to 0} \frac{f(a, b+h) - f(a, b)}{h}.$$

$\frac{\partial f}{\partial y}$ is called the **partial derivative** of f **with respect to** y. We also use the notation f_y in place of $\frac{\partial f}{\partial y}$.

The partial derivative with respect to y is calculated by treating the x variable as a constant.

Example 13. Let $f(x, y) = xy$. Then

$$\frac{\partial f}{\partial y}(x, y) = x, \quad \frac{\partial f}{\partial x}(x, y) = y.$$

Example 14. Let $f(x, y) = x^2 + 2xy^2$. Then

$$\frac{\partial f}{\partial x} = 2x + 2y^2, \quad \frac{\partial f}{\partial y} = 4xy.$$

Notice that we have used two different expressions for the partial derivatives in the above examples: $\frac{\partial f}{\partial y}(x, y)$ and $\frac{\partial f}{\partial y}$. The expression $\frac{\partial f}{\partial y}(x, y)$ is the correct one—i.e., it is the partial derivative with respect to y at the point (x, y). The expression $\frac{\partial f}{\partial y}$ is a shortened form of the same expression and is commonly used because notation can become extremely complicated very rapidly in the calculus of several variables. We **agree** from now on to use $\frac{\partial f}{\partial x}$ in place of $\frac{\partial f}{\partial x}(x, y)$, and similarly for $\frac{\partial f}{\partial y}(x, y)$ and $\frac{\partial f}{\partial y}$. Notice also that the letters or numbers occurring in the final expression **depend on the point** at which we are finding the partial derivative and not on the variables used in defining the function.

Example 15. Let $f(x, y) = x^2 + 2xy^2$

$$\frac{\partial f}{\partial x}(x, y) = \frac{\partial f}{\partial x} = 2x + 2y^2$$

$$\frac{\partial f}{\partial x}(a, b) = 2a + 2b^2$$

$$\frac{\partial f}{\partial x}(1, 2) = 2 + 8 = 10.$$

We summarize our knowledge of maxima and minima as follows: if f is a sufficiently regular function of two variables, defined on an open subset of $\mathbf{R}^2$, which has a maximum or minimum at (a, b), then

$$\frac{\partial f}{\partial x}(a, b) = \frac{\partial f}{\partial y}(a, b) = 0.$$

We tidy up our notation a little with the following definition.

Definition 16. *If f is a function of two variables defined on an open subset of $\mathbf{R}^2$ and $\dfrac{\partial f}{\partial x}$ and $\dfrac{\partial f}{\partial y}$ both exist let*

$$\operatorname{grad}(f) = \nabla f = \left(\frac{\partial f}{\partial x}, \frac{\partial f}{\partial y} \right).$$

A point (a, b) at which $\nabla f(a, b) = \left(\dfrac{\partial f}{\partial x}(a, b), \dfrac{\partial f}{\partial y}(a, b) \right) = (0, 0)$ is called a **critical point** *of f.*

The terminology $\operatorname{grad}(f)$ appears at first sight to be unrelated to the ideas we have been discussing. However, like most terminology it is not random, and a little investigation (or reflection) in this type of situation is often revealing. The term **grad** is a convenient way of writing **gradient**, which means **slope**. In the one variable case, the slope of the tangent line is the derivative and thus it is natural to use the term gradient in the above context. In some cases an understanding of the origin or the rationale for certain terminology can even help our understanding of the mathematics.

All maxima and minima are critical points. We can go further and say that all **local maxima and local minima** are critical points. A point (a, b) is a local maximum of f if there is a disc of positive radius about (a, b) such that $f(a, b) \geq f(x, y)$ for all (x, y) in this disc. Clearly local minima are defined in a similar fashion. Let us now see how this method works in practice. We consider the function $f(x, y) = x^2 + y^2$. We already know from our examination of level curves that f has a minimum at the origin. We have $\nabla f = (2x, 2y)$. At a critical point $\nabla f = (0, 0) = (2x, 2y)$. Hence $x = 0$ and $y = 0$ and so the origin is a critical point and the only point at which we can have a local maximum or minimum.

Critical Points

We were led to this method by examining the **cross sections** corresponding to $z = d$. We can also look at the cross sections $y = d$ and $x = d$ and see if they lead to anything new. The cross section $y = d$ means that we are considering the values of the function f when $y = d$, i.e., we are looking at $f(x, d)$. This, of course, leads to the partial derivative $\dfrac{\partial f}{\partial x}$ which we have already considered and the cross section $x = d$ leads to the function $f(d, y)$ and the partial derivative $\dfrac{\partial f}{\partial y}$. Thus we get no new information by considering these cross sections.

We arrived at the partial derivatives $\dfrac{\partial f}{\partial x}$ and $\dfrac{\partial f}{\partial y}$ by approaching the maximum along lines parallel to the x-axis and the y-axis. We may, of course, also approach the maximum along many other straight lines and it makes sense to see if these paths lead to new and useful methods. Our recently acquired experience suggests that we should encounter some kind of partial derivatives. We could follow the same method as before, but instead we try another approach—which is frequently used in mathematics. We rewrite what we have already obtained in another fashion. In this way we see immediately what the new derivatives will be.

We have
$$\frac{\partial f}{\partial x} = \lim_{h \to 0} \frac{f(x+h, y) - f(x, y)}{h}.$$

Now $(x + h, y) = (x, y) + (h, 0) = (x, y) + h(1, 0)$ and so
$$\frac{\partial f}{\partial x} = \lim_{h \to 0} \frac{f((x, y) + h(1, 0)) - f(x, y)}{h}.$$

Similarly
$$\frac{\partial f}{\partial y} = \lim_{h \to 0} \frac{f((x, y) + h(0, 1)) - f(x, y)}{h}.$$

We may thus consider $\dfrac{\partial f}{\partial x}$ and $\dfrac{\partial f}{\partial y}$ as partial derivatives in the directions $(1, 0)$ and $(0, 1)$. When we consider partial derivatives in this way we call them **directional derivatives**. We are thus led to the following general definition of directional derivative.

Definition 17. *If f is a function of the variables x and y and $\vec{v} = (v_1, v_2)$ is a vector in $\mathbf{R}^2$, then the directional derivative of f at (a, b) in the direction $\vec{v}$, $\dfrac{\partial f}{\partial \vec{v}}(a, b)$, is defined as*
$$\lim_{h \to 0} \frac{f((a, b) + h(v_1, v_2)) - f(a, b)}{h}.$$

If $e_1 = (1,0)$ and $e_2 = (0,1)$, then
$$\frac{\partial f}{\partial e_1} = \frac{\partial f}{\partial x} \quad \text{and} \quad \frac{\partial f}{\partial e_2} = \frac{\partial f}{\partial y}.$$

Since $\frac{\partial f}{\partial \vec{v}}$ measures the rate of change of f in the direction $\vec{v}$ this treats the vector $\vec{v}$ as a **unit of measurement** and thus there are good physical reasons for considering **unit vectors** only in Definition 17. Many authors take this approach and we were tempted to follow this route here. We resisted the temptation on mathematical grounds which, unfortunately, we do not have time to discuss here.

The arguments we have used previously can now be used to show the following: if f has a local maximum or a local minimum at (a,b), then $\frac{\partial f}{\partial \vec{v}}(a,b) = 0$ for **any** vector $\vec{v}$ in $\mathbf{R}^2$. We could, if we wished, work out $\frac{\partial f}{\partial \vec{v}}$ using the above definition, and indeed it is a worthwhile exercise, but instead we will use a formula—which we prove later—which enables us to calculate $\frac{\partial f}{\partial \vec{v}}$ in a very simple fashion.

If f is a sufficiently regular function, then
$$\frac{\partial f}{\partial \vec{v}} = v_1 \frac{\partial f}{\partial x} + v_2 \frac{\partial f}{\partial y} \qquad (*)$$

for any vector $\vec{v} = (v_1, v_2)$ in $\mathbf{R}^2$.

If $\vec{v} = (v_1, v_2)$ is a unit vector $(*)$ says that $\frac{\partial f}{\partial \vec{v}}$ is the **weighted average** of $\frac{\partial f}{\partial x}$ and $\frac{\partial f}{\partial y}$ with weights chosen proportionate to the coordinates of $\vec{v}$, v_1 and v_2. Such well-behaved functions must surely satisfy some regularity conditions.

Example 18. Let $f(x,y) = x \tan^{-1}(y)$ and let $\vec{v} = (2,3)$. Then
$$\frac{\partial f}{\partial x} = \tan^{-1}(y), \quad \frac{\partial f}{\partial y} = \frac{x}{1+y^2} \quad \text{and} \quad \frac{\partial f}{\partial \vec{v}} = 2\tan^{-1}(y) + \frac{3x}{1+y^2}.$$

We can rephrase this using the gradient and a type of multiplication of vectors in $\mathbf{R}^2$—the **dot product**, also called the **scalar product** and the **inner product**. This is defined as follows: if (x_1, y_1) and (x_2, y_2) are vectors in $\mathbf{R}^2$, then
$$(x_1, y_1) \cdot (x_2, y_2) = x_1 x_2 + y_1 y_2.$$

So the dot product is obtained by multiplying corresponding coordinates

Critical Points

and adding them together. We have, for $\vec{v} = (v_1, v_2)$,

$$\frac{\partial f}{\partial \vec{v}} = v_1 \frac{\partial f}{\partial x} + v_2 \frac{\partial f}{\partial y} = (v_1, v_2) \cdot \left(\frac{\partial f}{\partial x}, \frac{\partial f}{\partial y}\right)$$
$$= \vec{v} \cdot \nabla f.$$

We also use the notation $\langle \vec{v}, \vec{w} \rangle$ in place of $\vec{v} \cdot \vec{w}$, especially when $\vec{v}$ and $\vec{w}$ have complicated expressions. We recall two very important properties of the dot product which we shall have occasion to use in the near future.

(i) Two vectors are **perpendicular** if and only if their dot product is zero.

For lines with finite slope this is nothing other than the familiar formula that two non-vertical lines are perpendicular if and only if the product of their slopes is equal to -1. If (x_1, y_1) and (x_2, y_2) are non-vertical, then the slope m_i of the line through the origin determined by (x_i, y_i) is $\dfrac{y_i}{x_i}$ for $i = 1, 2$ (Figure 3.2).

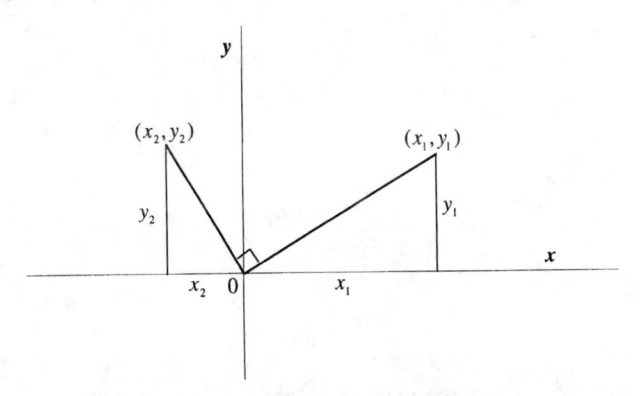

Figure 3.2

We have

$$m_1 \cdot m_2 = -1 \iff \frac{y_1}{x_1} \cdot \frac{y_2}{x_2} = -1$$
$$\iff y_1 y_2 = -x_1 x_2$$
$$\iff x_1 x_2 + y_1 y_2 = 0$$
$$\iff (x_1, y_1) \cdot (x_2, y_2) = 0.$$

(We use the notation $\iff$ to mean if and only if.)

(ii) The dot product of a vector with itself is equal to the square of its length.

This is just **Pythagoras' theorem** (Figure 3.3) since

$$(x_1, y_1) \cdot (x_1, y_1) = x_1^2 + y_1^2 = \|(x_1, y_1)\|^2.$$

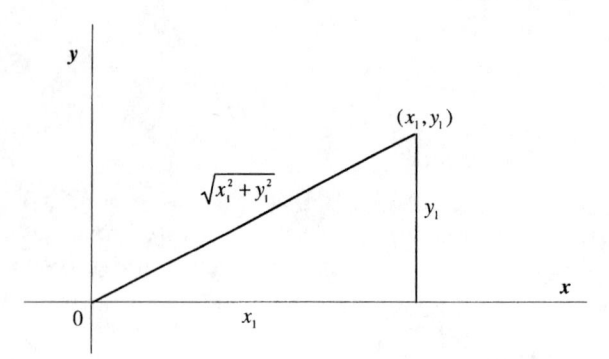

Figure 3.3

We use $\|(x_1, y_1)\|$ to denote the **length** of the vector (x_1, y_1).

If (a, b) is a critical point of f, then $\dfrac{\partial f}{\partial x}(a, b) = \dfrac{\partial f}{\partial y}(a, b) = 0$ and hence

$$\frac{\partial f}{\partial \vec{v}}(a, b) = v_1 \frac{\partial f}{\partial x}(a, b) + v_2 \frac{\partial f}{\partial y}(a, b) = 0.$$

Hence

$$\frac{\partial f}{\partial \vec{v}}(a, b) = 0$$

follows from $\dfrac{\partial f}{\partial x}(a, b) = 0$ and $\dfrac{\partial f}{\partial y}(a, b) = 0$ and we obtain no new information by putting the directional derivatives at (a, b) equal to 0.

Exercises

3.1 Find the directional derivatives of the functions given in Exercise 2.2 at the points $(0, 1)$, $(1, 2)$ and $(2, 1)$ in the directions $(-1, 0)$ and $(2, 3)$.

3.2 Find ∇f where f is any one of the functions given in Exercise 2.2.

3.3 Find the critical points of the functions

(a) $f(x, y) = x^3 + y^3 - 3xy$

(b) $g(x, y) = x^3 + 3xy^2 - 12x - 6y$

(c) $h(x, y) = x^4 + y^4 - 3x^3 - y^3$.

3.4 Find the length of the vector $(3, 4)$. Find all vectors of length 5 which are perpendicular to the vector $(3, 4)$.

3.5 If $f(x, y) = x^3 - 3xy^2$ and $g(x, y) = 3x^2y - y^3$ show that $\nabla f \cdot \nabla g = 0$.

4

Maxima and Minima

Summary. *We define second order partial and directional derivatives and the Hessian. We note that the mixed second order partial derivatives are often equal and obtain a practical criterion for locating some local maxima and minima within the set of critical points.*

Having found a set—the set of critical points—which includes all local maxima and minima, we turn back to the one-dimensional theory in order to see how to proceed. We recall the following fundamental result:

if f is twice differentiable on an open interval in $\mathbf{R}$, $f'(x_0) = 0$ and $f''(x_0) < 0$, then f has a local maximum at x_0.

This distinguishes certain critical points which are local maxima. Clearly, if a point (a, b) in $\mathbf{R}^2$ is a local maximum of a function of two variables, then from whatever direction we approach (a, b) it will be a local maximum and so we should now define second order directional derivatives and look at the critical points where all of these are negative.

Definition 19. *If f is a sufficiently regular function of two variables, then for any non-zero vectors $\vec{v}$ and $\vec{w}$ in $\mathbf{R}^2$ we let*

$$\frac{\partial^2 f}{\partial \vec{v} \partial \vec{w}} = \frac{\partial}{\partial \vec{v}}\left(\frac{\partial f}{\partial \vec{w}}\right).$$

If $\vec{v} = \vec{w}$ we write $\dfrac{\partial^2 f}{\partial \vec{v}^2}$. Taking $e_1 = (1,0)$ and $e_2 = (0,1)$ we define

$$\frac{\partial^2 f}{\partial e_1^2} = \frac{\partial}{\partial x}\left(\frac{\partial f}{\partial x}\right) = \frac{\partial^2 f}{\partial x^2} = f_{xx}$$

$$\frac{\partial^2 f}{\partial e_2 \partial e_1} = \frac{\partial}{\partial y}\left(\frac{\partial f}{\partial x}\right) = \frac{\partial^2 f}{\partial y \partial x} = f_{yx}$$

$$\frac{\partial^2 f}{\partial e_1 \partial e_2} = \frac{\partial}{\partial x}\left(\frac{\partial f}{\partial y}\right) = \frac{\partial^2 f}{\partial x \partial y} = f_{xy}$$

$$\frac{\partial^2 f}{\partial e_2^2} = \frac{\partial}{\partial y}\left(\frac{\partial f}{\partial y}\right) = \frac{\partial^2 f}{\partial y^2} = f_{yy}.$$

These four derivatives are called the **second order partial derivatives** of f. All the rules for differentiation, such as the product rule etc. are still valid. There is just one simple new rule to remember. While differentiating with respect to one variable you treat the other like a constant.

Example 20. Let $f(x,y) = xy e^x$

$$\frac{\partial f}{\partial x} = y(e^x + xe^x) , \quad \frac{\partial f}{\partial y} = xe^x$$

$$\frac{\partial^2 f}{\partial x^2} = \frac{\partial}{\partial x}\left(\frac{\partial f}{\partial x}\right) = y(e^x + e^x + xe^x) = y(2e^x + xe^x)$$

$$\frac{\partial^2 f}{\partial y \partial x} = \frac{\partial}{\partial y}\left(\frac{\partial f}{\partial x}\right) = \frac{\partial}{\partial y}(y(e^x + xe^x)) = e^x + xe^x$$

$$\frac{\partial^2 f}{\partial x \partial y} = \frac{\partial}{\partial x}\left(\frac{\partial f}{\partial y}\right) = \frac{\partial}{\partial x}(xe^x) = e^x + xe^x$$

$$\frac{\partial^2 f}{\partial y^2} = \frac{\partial}{\partial y}(xe^x) = 0.$$

The following proposition—which is the basis for our whole approach—appears plausible in view of the corresponding one-dimensional result, but a direct application of the one-dimensional theory, in case (i) for instance, only shows that we have a local maximum along **each line** through the critical point and it is just possible that a **non-linear** approach does not lead to a local maximum at the critical point (see Exercise 6.5). A somewhat similar and equally plausible result for continuous functions is false (Exercise 6.4). The result is, however, true and we provide a proof in Chapter 15.

Proposition 21. *If f is a sufficiently regular function defined on an open set in $\mathbf{R}^2$, then*

(i) f *has a local maximum at* (a,b) *if* $\nabla f(a,b) = (0,0)$ *and* $\dfrac{\partial^2 f}{\partial \vec{v}^2}(a,b) < 0$

Maxima and Minima

for all non-zero $\vec{v} \in \mathbf{R}^2$,

(ii) f has local minimum at (a,b) if $\nabla f(a,b) = (0,0)$ and $\dfrac{\partial^2 f}{\partial \vec{v}^2}(a,b) > 0$ for all non-zero $\vec{v} \in \mathbf{R}^2$.

The direct application of this proposition would involve a lot of work in practice—we would have to check $\dfrac{\partial^2 f}{\partial \vec{v}^2}$ for all $\vec{v}$ in $\mathbf{R}^2$—and so we will try to replace it by something more practical.

Before proceeding let us take one further look at the example we have just given. We notice that we are building up quite complicated formulae starting from a relatively simple function and any simplifications would clearly be in order. Such simplifications are often found by seeing if there are any remarkable coincidences which, on investigation, turn out to be examples of a general rule rather than mere coincidences. We notice two things. First $\dfrac{\partial^2 f}{\partial y^2} = 0$. Is this always the case? Well, if we interchange x and y in the above example then we get a new function—call it g—and $\dfrac{\partial^2 g}{\partial y^2} = x(2e^y + ye^y) \neq 0$. So this leads nowhere.

Second, we see that $\dfrac{\partial^2 f}{\partial y \partial x} = \dfrac{\partial^2 f}{\partial x \partial y}$. These derivatives are called—for obvious reasons—the **mixed second order partial derivatives**. To distinguish between them remember that we differentiate with respect to the variables on the right first. It is not **always** the case that these are equal but it is true **sufficiently often** to make it remarkable. We need f to be sufficiently regular. We have the following result.

Proposition 22. *If f is a continuous function defined on an open subset of $\mathbf{R}^2$ and $\dfrac{\partial^2 f}{\partial x^2}, \dfrac{\partial^2 f}{\partial y^2}, \dfrac{\partial^2 f}{\partial x \partial y}$ and $\dfrac{\partial^2 f}{\partial y \partial x}$ all exist and are all continuous, then*

$$\frac{\partial^2 f}{\partial x \partial y} = \frac{\partial^2 f}{\partial y \partial x}.$$

We will define continuous functions soon but will not prove this proposition. However, in Exercise 12.10 we do provide a simple rigorous proof which applies to almost all functions you will come across. This will not involve any complicated calculations and is very practical.

We now assume we are dealing with functions for which $\dfrac{\partial^2 f}{\partial x \partial y} = \dfrac{\partial^2 f}{\partial y \partial x}$.

Now let us return to working out $\dfrac{\partial^2 f}{\partial \vec{v}^2}$ for $\vec{v} = (v_1, v_2) \in \mathbf{R}^2$. We have

$$\frac{\partial^2 f}{\partial \vec{v}^2} = \frac{\partial}{\partial \vec{v}}\left(\frac{\partial f}{\partial \vec{v}}\right)$$

$$= \frac{\partial}{\partial \vec{v}}\left(v_1 \frac{\partial f}{\partial x} + v_2 \frac{\partial f}{\partial y}\right).$$

Let $g = v_1 \dfrac{\partial f}{\partial x} + v_2 \dfrac{\partial f}{\partial y}$. Then

$$\frac{\partial^2 f}{\partial \vec{v}^2} = \frac{\partial}{\partial \vec{v}}(g) = v_1 \frac{\partial g}{\partial x} + v_2 \frac{\partial g}{\partial y}. \qquad (*)$$

Since $g = v_1 \dfrac{\partial f}{\partial x} + v_2 \dfrac{\partial f}{\partial y}$ we have

$$\frac{\partial g}{\partial x} = v_1 \frac{\partial^2 f}{\partial x^2} + v_2 \frac{\partial^2 f}{\partial x \partial y}$$

and similarly $\dfrac{\partial g}{\partial y} = v_1 \dfrac{\partial^2 f}{\partial y \partial x} + v_2 \dfrac{\partial^2 f}{\partial y^2}$. Substituting these back into $(*)$, and using the assumption $\dfrac{\partial^2 f}{\partial x \partial y} = \dfrac{\partial^2 f}{\partial y \partial x}$, we get

$$\frac{\partial^2 f}{\partial \vec{v}^2} = v_1\left(v_1 \frac{\partial^2 f}{\partial x^2} + v_2 \frac{\partial^2 f}{\partial x \partial y}\right) + v_2\left(v_1 \frac{\partial^2 f}{\partial x \partial y} + v_2 \frac{\partial^2 f}{\partial y^2}\right)$$

$$= v_1^2 \frac{\partial^2 f}{\partial x^2} + 2v_1 v_2 \frac{\partial^2 f}{\partial x \partial y} + v_2^2 \frac{\partial^2 f}{\partial y^2}.$$

If $\vec{v} = (v_1, v_2)$, then using matrix notation and matrix multiplication we see that

$$\frac{\partial^2 f}{\partial \vec{v}^2} = (v_1, v_2) \begin{pmatrix} \dfrac{\partial^2 f}{\partial x^2} & \dfrac{\partial^2 f}{\partial x \partial y} \\ \dfrac{\partial^2 f}{\partial x \partial y} & \dfrac{\partial^2 f}{\partial y^2} \end{pmatrix} \begin{pmatrix} v_1 \\ v_2 \end{pmatrix}.$$

Definition 23. The **Hessian** of a sufficiently regular function f, H_f, is defined as

$$H_f = \begin{pmatrix} \dfrac{\partial^2 f}{\partial x^2} & \dfrac{\partial^2 f}{\partial x \partial y} \\ \dfrac{\partial^2 f}{\partial x \partial y} & \dfrac{\partial^2 f}{\partial y^2} \end{pmatrix}.$$

We will write ${}^t\vec{v}$ to denote the column vector $\begin{pmatrix} v_1 \\ v_2 \end{pmatrix}$ arising from the row

Maxima and Minima

vector $\vec{v} = (v_1, v_2)$ and call ${}^t\vec{v}$ the **transpose** of $\vec{v}$. Using Definition 23, matrix multiplication and the transpose we have

$$\frac{\partial^2 f}{\partial \vec{v}^2} = \vec{v} H_f \, {}^t\vec{v}.$$

Matrix theory is not essential for this book. We have included some matrix terminology mainly for those readers who are already familiar with linear algebra and we will give an example later showing how concepts from linear algebra arise naturally and are useful in differential calculus.

If we are at a critical point (a,b), then we must consider $\frac{\partial^2 f}{\partial \vec{v}^2}(a,b)$ which is, by the above,

$$v_1^2 \frac{\partial^2 f}{\partial x^2}(a,b) + 2 v_1 v_2 \frac{\partial^2 f}{\partial x \partial y}(a,b) + v_2^2 \frac{\partial^2 f}{\partial y^2}(a,b).$$

We want this to be strictly negative for all non-zero $\vec{v}$. Let us rewrite it as

$$v_1^2 A + 2 v_1 v_2 B + v_2^2 C$$

where $A = \frac{\partial^2 f}{\partial x^2}(a,b)$, $B = \frac{\partial^2 f}{\partial x \partial y}(a,b)$ and $C = \frac{\partial^2 f}{\partial y^2}(a,b)$. It becomes even more familiar if we rewrite it as $Ax^2 + 2Bxy + Cy^2$ where we have replaced v_1 by x and v_2 by y. This is a useful general principle—change the **unknowns** or **variables** to something familiar whenever you wish. Our experience in solving quadratics suggests that we try **completing squares**.

Let us recall how we solve the equation $ax^2 + bx + c = 0$. We have

$$ax^2 + bx + c = a(x^2 + \frac{b}{a}x + ?) + c - ?.$$

We get ? by taking half the coefficient of x and squaring it. So $? = \frac{b^2}{4a^2}$. We then have

$$ax^2 + bx + c = a\left(x^2 + \frac{b}{a}x + \frac{b^2}{4a^2}\right) + c - \frac{b^2}{4a^2} = 0,$$

i.e.,

$$a\left(x + \frac{b}{2a}\right)^2 = \frac{b^2}{4a} - c = \frac{b^2 - 4ac}{4a}.$$

Hence

$$\left(x + \frac{b}{2a}\right)^2 = \frac{b^2 - 4ac}{4a^2}$$

and

$$x + \frac{b}{2a} = \frac{\pm\sqrt{b^2 - 4ac}}{\sqrt{4a^2}} = \frac{\pm\sqrt{b^2 - 4ac}}{2a}.$$

So
$$x = -\frac{b}{2a} \pm \frac{\sqrt{b^2 - 4ac}}{2a} = \frac{-b \pm \sqrt{b^2 - 4ac}}{2a}$$
and this is the familiar pair of solutions of a quadratic equation.

In our case we apply the above method to an **inequality** and treat y like any other real number. If $Ax^2 + 2Bxy + Cy^2 < 0$ for all $(x,y) \neq (0,0)$, then taking (x,y), in turn, equal to $(1,0)$ and $(0,1)$ we obtain $A < 0$ and $C < 0$. We have

$$Ax^2 + 2Bxy + Cy^2 = A\left(x^2 + \frac{2B}{A}xy + \frac{B^2}{A^2}y^2\right) + Cy^2 - \frac{B^2}{A}y^2$$
$$= A\left(x + \frac{B}{A}y\right)^2 + \frac{1}{A}(AC - B^2)y^2. \qquad (1)$$

If $y = 1$ and $x = -B/A$, then (1) implies $AC - B^2 > 0$. Since

$$A\left(x + \frac{B}{A}y\right)^2 \leq 0$$

we have, by (1),

$$Ax^2 + 2Bxy + Cy^2 \leq \frac{1}{A}(AC - B^2)y^2 \qquad (2)$$

for all (x,y). If we replace x by y and A by C in Equation (2) we obtain

$$Ax^2 + 2Bxy + Cy^2 \leq \frac{1}{C}(AC - B^2)x^2. \qquad (3)$$

Let $\alpha = \frac{1}{2}(AC - B^2) \cdot \text{minimum}\,(\frac{1}{A}, \frac{1}{C})$. Clearly $\alpha < 0$. Adding (2) and (3) shows that condition (a) below implies (b). Clearly (b) implies (a). We have already noted that (a) implies (c) and (1) shows that (c) implies (a). We have thus established that the following conditions are equivalent:

(a) $Ax^2 + 2Bxy + Cy^2 < 0$ for all $(x,y) \neq (0,0)$,

(b) there exists $\alpha < 0$ such that $Ax^2 + 2Bxy + Cy^2 \leq \alpha(x^2 + y^2)$ for all $(x,y) \in \mathbf{R}^2$,

(c) $A < 0$ and $AC - B^2 > 0$.

It is important to note that the equivalence of these three conditions was fully established by completely elementary methods. Indeed, we did not use any differential calculus whatsoever in deriving it and we could just as easily have proved this equivalence in Chapter 1. The equivalence of (a) and (c) will be used in Chapter 15 to complete the proof of Proposition 21 and since this proposition is the main theoretical justification for our test for local maxima and minima, we have to be careful that we do not inadvertently

Maxima and Minima

fall into a circular argument, i.e., we have to make sure that Proposition 21 or any of its consequences—and there are many—are not used in the proof in Chapter 15. In Example 44 (Chapter 10) we shall give a more elegant proof that these conditions are equivalent using Lagrange multipliers.

We now use the equivalence of (a) and (c) to derive a practical test for locating local maxima. Since $AC-B^2$ is the determinant of the 2×2 **matrix** $\begin{pmatrix} A & B \\ B & C \end{pmatrix}$ we get, on replacing A, B and C by $\dfrac{\partial^2 f}{\partial x^2}(a,b)$, $\dfrac{\partial^2 f}{\partial x \partial y}(a,b)$ and $\dfrac{\partial^2 f}{\partial y^2}(a,b)$, respectively, the Hessian and a criterion for calculating local maxima.

Proposition 24. *If f is defined on an open subset of $\mathbf{R}^2$ and at (a,b) we have*

$$\begin{aligned}&\text{(i)} && \nabla f(a,b) = (0,0) \\ &\text{(ii)} && \det(H_{f(a,b)}) > 0 \\ &\text{(iii)} && \frac{\partial^2 f}{\partial x^2}(a,b) < 0\end{aligned}$$

then (a,b) is a local maximum of f.

This is a simple direct method of locating local maxima and is entirely in the spirit of the one-dimensional method. We arrived at Proposition 24 by a direct route guided only by the one-dimensional theory. We will give examples when we have worked out criteria for local minima.

Well, to find local minima we proceed in exactly the same fashion but we now want $\dfrac{\partial^2 f}{\partial \vec{v}^2} > 0$ for all non-zero $\vec{v}$ in $\mathbf{R}^2$. If we return to the identity (1) we see that

$$Ax^2 + 2Bxy + Cy^2 > 0$$

for all $(x,y) \neq (0,0)$ if and only if $A > 0$ and $AC - B^2 > 0$. Changing everything into partial derivatives and combining the results for local maxima and minima gives us the following proposition.

Proposition 25. *If f is defined on an open subset of $\mathbf{R}^2$ and at (a,b) we have $\nabla f(a,b) = (0,0)$ and $\det(H_{f(a,b)}) > 0$ then*

(a) *(a,b) is a **local maximum** if $\dfrac{\partial^2 f}{\partial x^2}(a,b) < 0$*

(b) *(a,b) is a **local minimum** if $\dfrac{\partial^2 f}{\partial x^2}(a,b) > 0$.*

So far, so good. We have found a method for locating local maxima and local minima. Are there any other possibilities?

Exercises

4.1 Find all the second order partial derivatives of
  (a) $z = xy^4 - 2x^2y^3 + 4x^2 - 3y$
  (b) $z = x^3 e^{-2y} + y^{-2} \cos x$
  (c) $z = y^2 \tan^{-1}(xy)$

and verify that $\dfrac{\partial^2 z}{\partial x \partial y} = \dfrac{\partial^2 z}{\partial y \partial x}$.

4.2 Find the critical points of the following functions:
  (a) $x^2 + 2xy + 3y^2$
  (b) $4x^3 + 4xy - y^2 - 4x$
  (c) $e^x \sin y$
  (d) $e^{2x+3y}(8x^2 - 6xy + 3y^2)$
  (e) $\sin x \sin y \sin(x+y)$
  (f) $xy + x^{-1} + y^{-1}$.

4.3 A function f defined on an open set in $\mathbf{R}^2$ is said to be **harmonic** if $\dfrac{\partial^2 f}{\partial x^2} + \dfrac{\partial^2 f}{\partial y^2} = 0$ at all points. Prove that the following functions are harmonic:
$$f(x,y) = \tan^{-1}\left(\frac{y}{x}\right), \quad g(x,y) = e^{-x}\cos y + e^{-y}\cos x.$$

4.4 Let $u(x,y) = e^x(x \cos y - y \sin y)$, $v(x,y) = e^x(y \cos y + x \sin y)$. Show that
  (i) $\dfrac{\partial u}{\partial x} = \dfrac{\partial v}{\partial y}$
  (ii) $\dfrac{\partial u}{\partial y} = -\dfrac{\partial v}{\partial x}$
  (iii) u and v are both harmonic.

4.5 By completing squares show that the level set
$$9x^2 + 25y^2 + 18x - 100y - 116 = 0$$
is an ellipse. Sketch the level set.

4.6 Let a and b be non-zero real numbers and let $c = ab$. Describe the following statements in terms of a and c.
  (i) a and b are both positive.
  (ii) a and b are both negative.
  (iii) a and b have different signs.

4.7 Obtain the test for locating local minima of f by applying Proposition 24 to $(-f)$.

5
Saddle Points

Summary. *In this chapter we encounter saddle points, that is to say critical points which yield a local maximum when approached from certain directions and a local minimum from other directions. Using the determinant of the Hessian we derive a method for deciding which non-degenerate critical points are local maxima, local minima and saddle points. Mathematical examples are given and we discuss how to find the least squares regression line.*

From the one-dimensional theory we know that Proposition 25 does not cover all possibilities. For example $f(x) = x^4$ has a local minimum at the origin but $f'(0) = f''(0) = 0$, while $g(x) = x^3$ has neither a local maximum nor a local minimum at the origin yet, $g'(0) = g''(0) = 0$ (Figure 5.1).

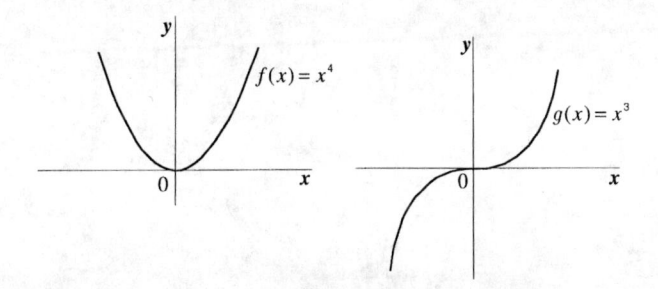

Figure 5.1

These examples show that if we have too many derivatives vanishing at a critical point, then we can say very little. In the two-dimensional case we consider the condition $\det(H_{f_{(a,b)}}) = 0$ at a critical point (a,b) to be the analogue of the vanishing of the second derivative at a critical point for a function of one variable. We call such points **degenerate critical points** and it is necessary to use other methods, which we will not discuss, to decide if a degenerate critical point is a local maximum or minimum or neither.

If $\det(H_f) \neq 0$ at the critical point (a,b), then we call the critical point **a non-degenerate critical point**. At non-degenerate critical points we have two possibilities, $\det(H_f) > 0$ or $\det(H_f) < 0$. If $\det(H_f) > 0$ then, using A, B and C as in the previous chapter, we have $AC - B^2 > 0$ and so $AC > B^2$. This means that A and C must either be both positive or both negative. By Proposition 24 if $A > 0$ then (a,b) is a local minimum and if $A < 0$ then (a,b) is a local maximum. So in fact we have already sorted out the $\det(H_f) > 0$ case.

Now suppose $\det(H_f) < 0$. We then have $AC - B^2 < 0$. Suppose $A > 0$. Then,

$$Ax^2 + 2Bxy + Cy^2 = A\left(x + \frac{B}{A}y\right)^2 + \frac{(AC - B^2)}{A}y^2.$$

When $x = 1$ and $y = 0$ we have $Ax^2 + 2Bxy + Cy^2 = A > 0$. When $x = -\frac{B}{A}$ and $y = 1$ we get

$$Ax^2 + 2Bxy + Cy^2 = \frac{AC - B^2}{A} < 0.$$

An examination of the other possibilities, $A < 0$ and $A = 0$, shows that we always have choices available such that $Ax^2 + 2Bxy + Cy^2$ can be made either positive or negative.

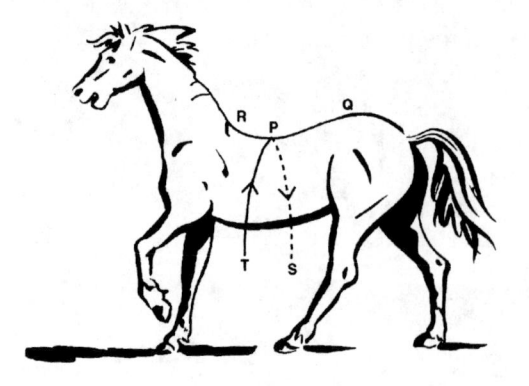

Figure 5.2

This means that, when $\det(H_{f(a,b)}) < 0$, there exists $\vec{v} \neq (0,0)$ such that $\frac{\partial^2 f}{\partial \vec{v}^2}(a,b) > 0$ and $\vec{w} \neq (0,0)$ such that $\frac{\partial^2 f}{\partial \vec{w}^2}(a,b) < 0$. We can only conclude from the one-dimensional theory that when we approach (a,b) in

Saddle Points

some direction we reach a local maximum and as we approach it from some other direction we achieve a local minimum.

Two classical cases where this occurs in nature are on the saddle of a horse and at a **mountain pass**. For this reason these points are called **saddle points**. Let us look closer at this phenomenon. Along the curve RPQ, in Figure 5.2, the point P is a minimum while along the curve TPS the point P is a maximum.

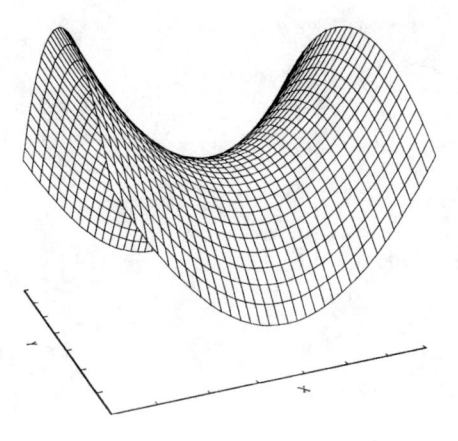

Figure 5.3

In Figure 5.3 we show part of the graph of the function $f(x,y) = x^2 - y^2$.

Saddle points cannot occur in the case of functions of one variable since we have only one cross section (the graph in $\mathbf{R}^2$) in this case and also we can only approach the point along one line (the x-axis). So a saddle point is a critical point in which some cross sections give a local maximum and others a local minimum. If $\det(H_{f(a,b)}) < 0$, then the critical point is a saddle point. We have now completed our analysis and summarize our results in the following theorem.

Theorem 26. *Let f be a sufficiently regular function defined on an open subset of $\mathbf{R}^2$. Points (a,b) for which $\nabla f(a,b) = (0,0)$ are called critical points.*

(a) *If (a,b) is a critical point and $\det(H_{f(a,b)}) > 0$ then, at (a,b), f has a local maximum if $\dfrac{\partial^2 f}{\partial x^2}(a,b) < 0$ and a local minimum if $\dfrac{\partial^2 f}{\partial x^2}(a,b) > 0$.*

(b) *If (a,b) is a critical point and $\det(H_{f(a,b)}) < 0$ then (a,b) is a saddle point.*

(c) *If (a,b) is a critical point and $\det(H_{f(a,b)}) = 0$ then (a,b) is a degenerate critical point and other methods are required in order to determine whether it is a local maximum, a local minimum or a saddle point.*

Example 27. Let $f(x,y) = x^2 - 3xy - y^2 - 2y - 6x$. Then
$$\nabla f = (2x - 3y - 6, -3x - 2y - 2).$$
To find the critical points we put both of the entries in ∇f equal to zero.

We obtain two linear equations
$$2x - 3y = 6$$
$$3x + 2y = -2$$
which we solve in the usual fashion
$$6x - 9y = 18$$
$$6x + 4y = -4$$
$$-13y = 22$$
$y = -\frac{22}{13}$, $2x = 6 + 3y = 6 - \frac{66}{13} = \frac{78-66}{13} = \frac{12}{13}$.
So $\left(\frac{6}{13}, -\frac{22}{13}\right)$ is the only critical point.
We also have
$$\frac{\partial^2 f}{\partial x^2} = 2, \qquad \frac{\partial^2 f}{\partial x \partial y} = -3, \qquad \frac{\partial^2 f}{\partial y^2} = -2$$
and
$$H_{f\left(\frac{6}{13}, -\frac{22}{13}\right)} = \begin{pmatrix} 2 & -3 \\ -3 & -2 \end{pmatrix}.$$
Since $\det(H_f) = -4 - 9 = -13 < 0$ the point $\left(\dfrac{6}{13}, -\dfrac{22}{13}\right)$ is a saddle point.

Example 28. Let $f(x,y) = (x-1)e^{xy}$. Then
$$\nabla f = \left(\frac{\partial f}{\partial x}, \frac{\partial f}{\partial y}\right) = \left(e^{xy} + (x-1)ye^{xy}, (x-1)xe^{xy}\right)$$
$$= \left(e^{xy}(1 + xy - y), (x-1)xe^{xy}\right).$$
To find the critical points we must solve
$$e^{xy}(1 + xy - y) = 0 \quad \text{and} \quad (x-1)xe^{xy} = 0.$$
There is no typical set of equations which arise in this type of situation and the same is true when we consider Lagrange multipliers later and so we have no method that can be applied in all cases to get solutions. It is usually a help to start with the simplest looking equation and to use it to eliminate one unknown and then to substitute into the next simplest equation and so on. In these situation do not be afraid to experiment and to try different approaches.

Saddle Points

In this case the simplest looking equation is the second one $(x-1)xe^{xy} = 0$. As the exponential function is never zero we may divide both sides by e^{xy} to obtain $x(x-1) = 0$. If two numbers are multiplied together to give 0, then at least one of them must be zero—we use this over and over again—so that we have either $x = 0$ or $x = 1$.

If $x = 0$ then $e^{xy}(1+xy-y) = 0$ becomes $e^0(1+0-y) = 0$, i.e., $1-y = 0$ and $y = 1$ so that one critical point is the point $(0, 1)$.

If $x = 1$ then $e^{xy}(1+xy-y) = 0$ becomes $e^y(1+y-y) = 0$, i.e., $e^y \cdot 1 = 0$. Since $e^y \neq 0$, no matter what the value of y, we have arrived at something that is impossible and so we cannot have a solution in this case.

We have just one critical point $(0, 1)$ (Figure 5.4).

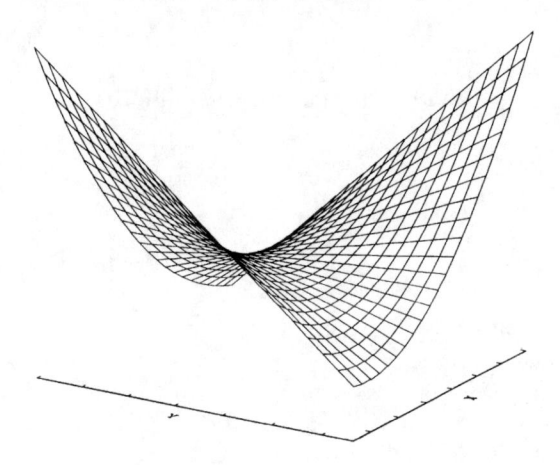

Figure 5.4

Now
$$\frac{\partial^2 f}{\partial x^2} = \frac{\partial}{\partial x}\left(e^{xy}(1+xy-y)\right)$$
$$= ye^{xy}(1+xy-y) + ye^{xy}$$

and $\dfrac{\partial^2 f}{\partial x^2}(0, 1) = 1e^0(1+0-1) + e^0 \cdot 1 = 1$. Also

$$\frac{\partial^2 f}{\partial y^2} = \frac{\partial}{\partial y}\left((x-1)xe^{xy}\right) = (x-1)xe^{xy} \cdot x$$

and $\dfrac{\partial^2 f}{\partial y^2}(0, 1) = (0-1) \cdot 0 \cdot e^0 \cdot 0 = 0$

$$\frac{\partial^2 f}{\partial x \partial y} = \frac{\partial^2 f}{\partial y \partial x} = \frac{\partial}{\partial y}\left(e^{xy}(1+xy-y)\right) = xe^{xy}(1+xy-y) + e^{xy}(x-1).$$

So $\dfrac{\partial^2 f}{\partial x \partial y}(0, 1) = 0 \cdot e^0(1+0 \cdot 1-1) + e^0(0-1) = -1$.

We have
$$H_{f(0,1)} = \begin{pmatrix} 1 & -1 \\ -1 & 0 \end{pmatrix} \quad \text{and} \quad \det(H_{f(0,1)}) = -1 < 0$$
and $(0,1)$ is a saddle point of the function f.

Example 29. Let $f(x,y) = x^3 + y^3 - 3xy$. Then
$$\nabla f = (3x^2 - 3y, 3y^2 - 3x).$$
For critical points we must solve
$$3x^2 - 3y = 0$$
$$3y^2 - 3x = 0.$$
From the first equation $y = x^2$. Substituting into the second equation gives
$$(x^2)^2 = x, \quad \text{i.e., } x^4 = x \text{ so } x^4 - x = 0.$$
Hence $x(x^3 - 1) = 0$ and $x = 0$ or $x^3 = 1$, i.e., $x = 1$.

If $x = 1$ then $y = x^2 = 1$ and if $x = 0$ then $y = 0$. So our critical points are $(0,0)$ and $(1,1)$.

Notice that our original function was **symmetric** with respect to the variables x and y, in other words $f(x,y) = f(y,x)$, and that we obtained critical points which were also symmetric. If we interchange the coordinates in $(0,0)$ and $(1,1)$ we get the same points. In fact it is easy to see that if (a,b) is a critical point of the symmetric function f, then (b,a) is also a critical point. However, it is not always the case that (a,b) being a local maximum implies that (b,a) is also a local maximum. This lack of symmetry does not occur often and if you come across it you should check your calculations.

Since
$$H_f = \begin{pmatrix} 6x & -3 \\ -3 & 6y \end{pmatrix}$$
we have
$$\det(H_{f(0,0)}) = \det\begin{pmatrix} 0 & -3 \\ -3 & 0 \end{pmatrix} = -9 < 0$$
and $(0,0)$ is a saddle point. Also,
$$\det(H_{f(1,1)}) = \det\begin{pmatrix} 6 & -3 \\ -3 & 6 \end{pmatrix} = 36 - 9 = 27 > 0$$
so we have either a local maximum or minimum at $(1,1)$. Since $\dfrac{\partial^2 f}{\partial x^2}(1,1) = 6 > 0$ it follows that f has a local minimum at $(1,1)$.

Saddle Points

The test for finding maxima and minima is frequently used in the application of mathematics to real life situations such as economics, statistics, physics and engineering. A set of mathematical equations which claims to describe a practical situation is called a **mathematical model**. Initially the set of equations in the model is based on experiments or previous experience of similar situations. A mathematical analysis of the model is used to derive new results which are in turn tested in the practical situation and the accuracy of the results is then used to see if the model is realistic or if it needs to be improved by adding new equations or adjusting the original ones. In this way quite complicated models can be gradually constructed.

An example occurs in examining the results $(x_1, y_1), \ldots, (x_n, y_n)$ of n experiments where x_i is, for instance, the amount of fertilizer used on a crop and y_i is the yield. An initial mathematical model of the relationship between x and y is often obtained by considering y as a linear function of x, i.e.,

$$y = ax + b$$

and the problem is to decide from the experiments which numbers a and b give the "best" linear function, i.e., the best linear model.

The word "best" is very subjective and a decision has to be taken. The **least squares estimate** is arrived at by choosing a and b so as to minimize the sum

$$E(a,b) = \sum_{i=1}^{n} (ax_i + b - y_i)^2$$

and the resulting line, which is often used in statistics, is called the **least squares regression line**.

To find this line we apply our method and first obtain the critical points $\dfrac{\partial E}{\partial a} = \dfrac{\partial E}{\partial b} = 0$. The only critical point is given by

$$a = \frac{n \sum_{i=1}^{n} x_i y_i - \sum_{i=1}^{n} x_i \cdot \sum_{i=1}^{n} y_i}{n \sum_{i=1}^{n} x_i^2 - \left(\sum_{i=1}^{n} x_i\right)^2}$$

and

$$b = \frac{1}{n}\left(\sum_{i=1}^{n} y_i - a \sum_{i=1}^{n} x_i\right)$$

and an application of the Hessian shows that E has a local minimum at this point. Since $E(a,b)$ tends to infinity as $\|(a,b)\|$ tends to infinity, it follows that the local minimum is an absolute minimum. See also Exercise 1.7 (i).

Exercises

5.1 Find all non-degenerate critical points of the functions given in Exercises 3.3 and 4.2 and determine which are local maxima, local minima and saddle points.

5.2 Find all local maxima, local minima and saddle points of the function $\sin x + \sin y + \sin(x+y)$ inside the rectangle $(0, \pi) \times (0, \pi)$.

5.3 Find the least squares regression line for the data

Fertilizer (x)	1.0	2.0	3.0	4.0	5.0
Yield (y)	28	40	47	52	55

Plot the points and the least squares regression line on the same diagram. Do you feel that the linear model is appropriate? Can you suggest a better model?

5.4 A firm sells two goods G_1 and G_2 at £1000 and £800, respectively. The total cost of producing x items of G_1 and y items of G_2 is
$$2x^2 + 2xy + y^2.$$
Find the output that maximises the profit and the maximum profit.

5.5 Find the maximum and minimum of
$$f(x,y) = 2xy - (1 - x^2 - y^2)^{3/2}$$
on the set $\{(x,y); x^2 + y^2 \leq 1\}$.

5.6 Consider the function $f(x,y) = \sqrt{x^2 + y^2} + \sqrt{(x-1)^2 + y^2}$. Explain why this function must have an absolute minimum at some point in $\mathbf{R}^2$. Give reasons why this minimum must occur at a point with zero y-coordinate. Having eliminated y find, without using the differential calculus, the absolute minimum of f and the points at which it is acheived. Show that the partial derivatives do not exist at all points where the minimum is achieved. Sketch this problem (not the function) using distances and solve it geometrically.

6

Sufficiently Regular Functions

Summary. *We recall the assumptions required of functions in order to apply the method of finding maxima and minima. We define limits and continuity and obtain a practical one-dimensional method of identifying large collections of functions of two variables which are sufficiently regular for our purposes. We state the fundamental theorem on the existence of maxima and minima.*

We have found a simple and efficient method of locating local maxima and minima for functions on an open subset of $\mathbf{R}^2$. In arriving at this method we had to assume on a number of occasions that our functions were **sufficiently regular**. We must now be a little more precise about this, otherwise we will not be sure if we are applying the method in suitable situations.

An examination of our approach shows that the following regularity conditions are sufficient for our purposes:

(a) *existence of first and second order partial derivatives,*
(b) *continuity of first order partial derivatives in order to express the first order directional derivatives in terms of first order partial derivatives,*
(c) *continuity of second order partial derivatives in order to have the mixed second order partial derivatives equal, to express the second order directional derivatives in terms of the second order partial derivatives and to prove Proposition 21.*

It is interesting to note that only (a) appears formally in our method and that (b) and (c) are used to derive the method. If functions satisfy the conditions in (c), then clearly they also satisfy (a) and (b) and so we can apply Theorem 26 to a function f whenever the first and second order partial derivatives of f exist and are continuous.

Since continuity and derivatives are both defined using the more basic concept of **limit** we begin by studying this notion. Limits and continuity are, as we shall see later in this chapter, rather subtle concepts.

Definition 30. *If f is defined on a subset U of $\mathbf{R}^2$ and $(a, b) \in U$ then*

$$\lim_{(x,y) \to (a,b)} f(x,y) = L$$

if for each positive number ϵ there exists a positive number δ such that

$$|f(x,y) - L| < \epsilon$$

whenever $(x, y) \in U$ and $(x - a)^2 + (y - b)^2 < \delta^2$.

In other words $f(x, y)$ is close to L when (x, y) is close to (a, b). In particular, if we move along any curve which ends at (a, b), then the values of f along the curve will get close to L. In terms of sequences we can say the following: if $(x_n)_n$ is a sequence which tends to a and $(y_n)_n$ is a sequence which tends to b then $(x_n, y_n)_n$ is a sequence which tends to (a, b) and so $\lim_{n \to \infty} f(x_n, y_n) = L$. The converse is true and both curves and sequences can be used to define limits; if the limit along **every curve** or along **every sequence** exists and gives the same value L, then the limit exists and equals L. We must have the **same limit** when we arrive at (a, b) from **any direction**.

Definition 31. *If f is defined on a subset U of $\mathbf{R}^2$, then f is continuous on U if*

$$\lim_{(x,y) \to (x_0,y_0)} f(x,y) = f(x_0, y_0)$$

for any point (x_0, y_0) in U.

The sum, difference, product, quotient, composition of limits and continuous functions behave as in the one-dimensional case and so the sum or product or composition of continuous functions is continuous and so is the quotient as long as we do not divide by zero. We now see that many of the functions of two variables that arise in practice come from the one-dimensional theory, and we can use continuity and differentiability properties in the one-dimensional case to show that many functions of two variables are continuous and differentiable.

The general principle is the following. If $f \colon (a, b) \subset \mathbf{R} \longrightarrow \mathbf{R}$ is continuous, then f can be identified with a continuous function of two variables, $\widehat{f}$, in the following way. Let

$$\widehat{f} \colon (a, b) \times \mathbf{R} \longrightarrow \mathbf{R} \text{ be given by } \widehat{f}(x, y) = f(x).$$

Sufficiently Regular Functions

If (x, y) is close to (x_0, y_0), then x is close to x_0 and so $f(x)$ is close to $f(x_0)$ (since f is continuous on (a, b)) and so $\widehat{f}(x, y) = f(x)$ is close to $\widehat{f}(x_0, y_0) = f(x_0)$. Hence $\widehat{f}$ is continuous.

Now if we consider a function such as

$$(x, y) \in \mathbf{R}^2 \longrightarrow \sin(xy)$$

then the functions $g(x, y) = x$ and $h(x, y) = y$ are both continuous. Since the product of continuous functions is easily seen to be continuous it follows that

$$gh(x, y) = g(x, y)h(x, y) = xy$$

is continuous and now if we compose it with the sine function, which we know to be a continuous function from $\mathbf{R}$ into $\mathbf{R}$, we see that

$$xy \longrightarrow \sin(xy)$$

is continuous.

Most, if not all, of the functions of two variables that we encounter can be treated in this fashion and so will be continuous and it will not be necessary to look at limits in $\mathbf{R}^2$. Moreover, if we start with functions on $\mathbf{R}$ that have one, two or more derivatives (as functions of one variable) then the functions that we end up with will have the same degree of partial differentiability.

This follows, using the same notation, since

$$\frac{\widehat{f}(x + \Delta x, y) - \widehat{f}(x, y)}{\Delta x} = \frac{f(x + \Delta x) - f(x)}{\Delta x}.$$

Taking limits we get

$$\frac{\partial \widehat{f}}{\partial x} = f'(x).$$

Similarly $\dfrac{\partial \widehat{f}}{\partial y} = 0$ and $\dfrac{\partial^2 \widehat{f}}{\partial x^2} = f''(x)$.

So, for instance, we have already seen that the function $\sin(xy)$ is obtained from functions which have derivatives of all orders and that all these derivatives are continuous. Hence $f(x, y) = \sin(xy)$ has partial derivatives of **all** orders and all are continuous, and we conclude that $\dfrac{\partial^2 f}{\partial x \partial y} = \dfrac{\partial^2 f}{\partial y \partial x}$ in this case and our method for finding maxima and minima applies to this function. A look over the examples we have given up to now shows, in fact, that all the functions we have considered have partial derivatives of all orders and that these partial derivatives are all continuous. To find functions which are not sufficiently regular is quite difficult.

Now we look at some limits in two dimensions which **cannot** be reduced to the one-dimensional setting. We will not prove or give any general results but instead look at a number of standard examples. From the one-dimensional theory we know that limits involving continuous functions do not cause any difficulty unless we have a situation like the following:

$$\lim_{x \to 0} \frac{x \sin x}{e^x - 1 - x}.$$

The functions $x \sin x$ and $e^x - 1 - x$ are both differentiable infinitely often but both take the value 0 at the origin so substituting in the limit values at the origin gives $\frac{0}{0}$ and there is no fixed meaning for this expression. The important feature in such a case is **the rate at which they both go to zero** and since derivatives involve rates of change it is not surprising (with hindsight of course) that one of the main methods for evaluating such limits—**l'Hôpital's rule**—involves derivatives. In the two-dimensional case we have no such simple rule so instead we consider a few typical examples.

Example 32.
$$\lim_{(x,y) \to (0,0)} \frac{x^4 + xy^2 + yx^3 + y^3}{x^3 + y^2}.$$

If we let $(x, y) = (0, 0)$ we get $\frac{0}{0}$. Now

$$\frac{x^4 + xy^2 + yx^3 + y^3}{x^3 + y^2} = \frac{x^4 + x^3 y + xy^2 + y^3}{x^3 + y^2}$$

$$= \frac{x^3(x + y) + y^2(x + y)}{x^3 + y^2}$$

$$= \frac{(x^3 + y^2)(x + y)}{x^3 + y^2}$$

$$= x + y$$

and

$$\lim_{(x,y) \to (0,0)} \frac{x^4 + xy^2 + yx^3 + y^3}{x^3 + y^2} = \lim_{(x,y) \to (0,0)} x + y = 0.$$

The key to this solution is to divide the denominator into the numerator. To see how to do this, notice that there is an x^3 term in the denominator and the first step is to bring together the terms in the numerator which contain an x^3 and then do the same with y^2. After factoring out x^3 and y^2 we see that $x + y$ is multiplied by both and now it is clear how to proceed. This simple approach often works.

Example 33. $\lim\limits_{(x,y) \to (0,0)} \dfrac{x^2 - y^2}{x^2 + y^2}$. Clearly there is no way of dividing the numerator into the denominator. Let us approach $(0,0)$ along different

Sufficiently Regular Functions

lines. As with finding the maximum and minimum points we take an east–west and then a north–south approach (Figure 6.1).

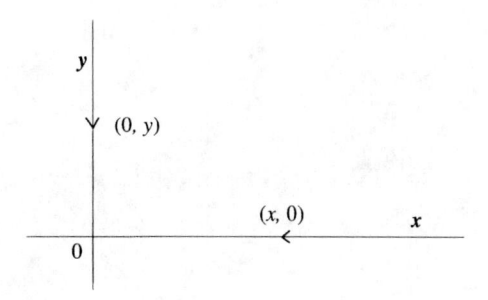

Figure 6.1

If we approach in an east–west fashion, then $y = 0$ and $\dfrac{x^2 - y^2}{x^2 + y^2} = \dfrac{x^2}{x^2} = 1$ and the limit along this line is 1. If we approach in a north–south fashion, then $x = 0$ and $\dfrac{x^2 - y^2}{x^2 + y^2} = \dfrac{-y^2}{y^2} = -1$ and the limit is -1. We conclude that $\lim\limits_{(x,y) \to (0,0)} \dfrac{x^2 - y^2}{x^2 + y^2}$ does not exist since we have found **two different limits** by approaching the origin along two different lines.

Example 34. $\lim\limits_{(x,y) \to (0,0)} \dfrac{x^2 y}{x^4 + y^2}$. Along the east–west line we have

$$\frac{x^2 y}{x^4 + y^2} = \frac{0}{x^4} \longrightarrow 0 \quad \text{as } x \to 0.$$

Along the north–south line we have

$$\frac{x^2 y}{x^4 + y^2} = \frac{0}{y^2} \longrightarrow 0 \quad \text{as } y \to 0.$$

So we get the same limit along both lines. We try an arbitrary line which goes through the origin. If this line has slope m, then the line has equation $y = mx$ and we are considering the points (x, mx). Along this line we have

$$\frac{x^2 y}{x^4 + y^2} = \frac{x^2 \cdot mx}{x^4 + m^2 x^2} = \frac{mx}{x^2 + m^2} \longrightarrow \frac{0}{m^2} = 0 \quad \text{as } x \to 0.$$

So we conclude that

$$\frac{x^2 y}{x^4 + y^2} \longrightarrow 0 \quad \text{as } (x, y) \to (0, 0)$$

along **any** straight line through the origin.

To show that $\lim_{(x,y)\to(0,0)} \dfrac{x^2 y}{x^4 + y^2} = 0$ we must show that $\dfrac{x^2 y}{x^4 + y^2} \to 0$ along **any** curve which passes through the origin. Since there are many different curves we cannot hope to consider them all. Let us, however, try another simple curve which is not a straight line. We consider the curve $y = x^2$ (Figure 6.2).

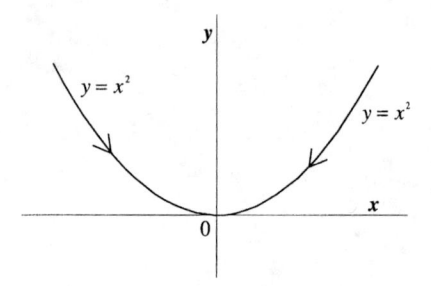

Figure 6.2

It is not a straight line and it does pass through the origin. On this curve $y = x^2$ and so

$$\frac{x^2 y}{x^4 + y^2} = \frac{x^2 x^2}{x^4 + (x^2)^2} = \frac{x^4}{x^4 + x^4} = \frac{1}{2}$$

and so we get a limit which is not zero and conclude that $\lim_{(x,y)\to(0,0)} \dfrac{x^2 y}{x^4 + y^2}$ does not exist.

Just out of curiosity, let us consider the limit along a more complicated curve, say $y = x^4$. Then

$$\frac{x^2 y}{x^4 + y^2} = \frac{x^2 x^4}{x^4 + x^8} = \frac{x^6}{x^4 + x^8} = \frac{x^2}{1 + x^4} \longrightarrow 0.$$

Along $y = x^5$

$$\frac{x^2 y}{x^4 + y^2} = \frac{x^2 x^5}{x^4 + x^{10}} = \frac{x^7}{x^4 + x^{10}} = \frac{x^3}{1 + x^6} \longrightarrow 0.$$

This may be a little surprising, but if we think about it we are looking at our function along curves which are getting progressively flatter as we take higher and higher powers and so they are getting closer to the x-axis which is a straight line (Figure 6.3) and along any line $\dfrac{x^2 y}{x^3 + y^2}$ tends to zero.

If you experiment with functions of the above kind and vary the powers you can construct a function which goes to zero along any curve of the form $y = x^j$ for $j \neq m$ but does not go to zero along $y = x^m$.

Sufficiently Regular Functions 43

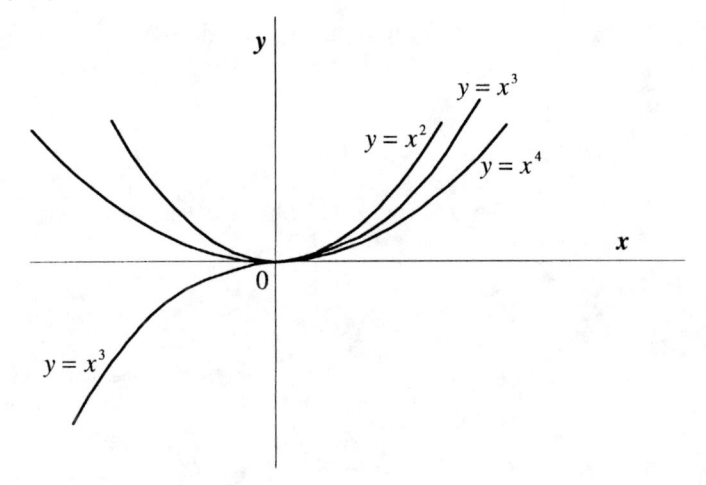

Figure 6.3

We have not established any general results but have seen two possible methods of approach to such limits—try to divide the numerator into the denominator and if this fails then evaluate the function along various curves which go through the origin.

Limits and sequences also play a role in establishing an important theoretical result with useful practical implications. To motivate this result we return to our original problem—to find the maximum and minimum of a function f on a set A. Critical points and, later, Lagrange multipliers help us to identify a relatively small subset B of A which contains all local maxima and minima. If we **know** that f has a maximum and minimum on A then these can be found by inspecting the values of f on the more managable set B. Without this knowledge we may well be looking for something that does not exist. So it is useful to know **in advance** that the function f does have a maximum and a minimum on the set A. The **fundamental existence theorem** for maxima and minima, which generalises Theorem 1, provides us with this advance knowledge. We must place conditions on both f and A.

A subset A of $\mathbf{R}^2$ is
(i) **bounded** if there exists M such that $\|(x,y)\| \leq M$ for all (x,y) in A,
(ii) **closed** if its complement is open.

The fundamental existence theorem states that *a continuous function on a closed bounded subset of* $\mathbf{R}^2$ *has a maximum and a minimum.*

In many cases we are given an open set U and are interested in finding the maximum and minimum of a function on the closure $\overline{U}$ of U, where

$$\overline{U} = \{(x,y) \in \mathbf{R}^2; \text{ there exists } (x_n, y_n)_n, \text{ a sequence in } U, \text{ such that } (x_n, y_n) \to (x,y) \text{ as } n \to \infty\}.$$

The set $\overline{U} \backslash U$ is the **boundary** of U. The set $\overline{U}$ is always closed and is bounded if and only if U is bounded.

Exercises

6.1 Let $f(x,y) = x$ and $g(x,y) = y$ for all $x, y \in \mathbf{R}$. Express the function

$$\sin^2\left(\frac{xy}{x^2 + y^2 + 1}\right)$$

in terms of f and g and functions from $\mathbf{R}$ into $\mathbf{R}$.

6.2 Find the limit or prove it does not exist in each of the following cases.

(a) $\displaystyle\lim_{(x,y) \to (0,0)} \frac{x^2 - 2}{3 + xy}$

(b) $\displaystyle\lim_{(x,y) \to (0,0)} \frac{x^3 - x^2y + xy^2 - y^3}{x^2 + y^2}$

(c) $\displaystyle\lim_{(x,y) \to (0,0)} \frac{x^2 - 2xy + 5y^2}{3x^2 + 4y^2}$

(d) $\displaystyle\lim_{(x,y) \to (1,2)} \frac{xy - 2x - y + 2}{x^2 + y^2 - 2x - 4y + 5}$

(e) $\displaystyle\lim_{(x,y) \to (0,0)} \frac{x^4 - y^4}{x^2 + y^2}$

(f) $\displaystyle\lim_{(x,y) \to (0,0)} \frac{x^4 - y^4}{x^2 - y^2}$.

6.3 Show that the function

$$f(x,y) = \frac{x^3 y^3}{x^{12} + y^4}$$

tends to zero along any curve of the form $y = x^m$, $m \neq 3$, but does not tend to zero along the curve $y = x^3$.

6.4 Let

$$f(x,y) = \begin{cases} 0 & \text{if } (x,y) = (0,0) \\ \dfrac{x^2 y}{x^4 + y^2} & \text{if } (x,y) \neq (0,0). \end{cases}$$

Show that f is not continuous at $(0,0)$ but that f is continuous when restricted to any straight line through the origin.

6.5 Let $f(x,y) = (y - x^2)(y - 3x^2)$. Show that the restriction of f to any straight line through the origin has a local minimum at the origin while the restriction of f to the curve $y = 2x^2$ has a local maximum at the origin.

7

Linear Approximation

Summary. *We linearly approximate functions which have continuous first order partial derivatives and using this approximation provide a complete proof of the standard method of finding directional derivatives.*

To be absolutely sure that our method of finding maxima and minima is correct, not only should we check that we are applying the method to the correct set of functions, but we should also prove that the results used to derive the method are true. This amounts to proving the following three results.

(a) *If f is a continuous function of two variables and $\dfrac{\partial f}{\partial x}$ and $\dfrac{\partial f}{\partial y}$ both exist and are continuous, then for any vector $\vec{v} = (v_1, v_2)$ in $\mathbf{R}^2$*
$$\frac{\partial f}{\partial \vec{v}} = v_1 \frac{\partial f}{\partial x} + v_2 \frac{\partial f}{\partial y}.$$

(b) *If all the second order partial derivatives of f exist and are continuous, then*
$$\frac{\partial^2 f}{\partial x \partial y} = \frac{\partial^2 f}{\partial y \partial x}.$$

(c) *If all the second order partial derivatives of f exist and are continuous, then f satisfies the hypothesis of Proposition 21.*

We shall prove (a) and (c). We do not prove (b) but outline, in Exercise 12.9, a simple proof which is valid for most functions you are likely to encounter. We prove (c) in Chapter 15. To begin our proof of (a) we reinterpret the derivative as an approximation. To orient ourselves we return once more to the one-dimensional situation. If f is a differentiable function of one variable, then

$$f'(x) = \lim_{\Delta x \to 0} \frac{f(x + \Delta x) - f(x)}{\Delta x}.$$

Hence

$$\frac{f(x + \Delta x) - f(x)}{\Delta x} = f'(x) + \text{``something small''}$$

and

$$f(x + \Delta x) = f(x) + f'(x) \cdot \Delta x + \text{``something small''} \cdot \Delta x.$$

This "something small" depends on both x and Δx and so we may write it as a function, which we will call g, of the variables x and Δx. So

$$f(x + \Delta x) = f(x) + f'(x) \cdot \Delta x + g(x, \Delta x) \cdot \Delta x$$

and $g(x, \Delta x) \to 0$ as $\Delta x \to 0$.

Do not forget when taking such limits that x is a fixed point and the formula makes sense as long as $f(x + \Delta x)$ makes sense. The only thing that varies is Δx and we are only interested in the case where Δx is small.

We may regard $f(x) + f'(x) \cdot \Delta x$ as an **approximation** of $f(x + \Delta x)$ and, as in any approximation, the absolute value of the difference between the true value, $f(x + \Delta x)$, and the approximation, $f(x) + f'(x) \cdot \Delta x$, is called the **error**. So our error is

$$\left| f(x + \Delta x) - f(x) - f'(x) \cdot \Delta x \right| = \left| g(x, \Delta x) \right| \left| \Delta x \right|$$

and we know $\left| g(x, \Delta x) \right| \to 0$ as $\Delta x \to 0$. Graphically we can represent this approximation as shown in Figure 7.1.

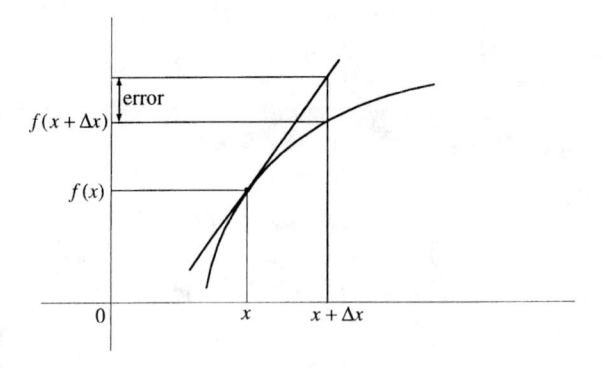

Figure 7.1

Linear Approximation

We call $f(x) + f'(x) \cdot \Delta x$ a **linear** approximation since its graph, with variable Δx, is a **straight line**. It is the **unique** linear approximation to $f(x + \Delta x)$ with the same quality of error, for suppose $A + B \cdot \Delta x$ is an approximation to $f(x+\Delta x)$ with error $h(x, \Delta x) \cdot \Delta x$ satisfying $h(x, \Delta x) \to 0$ as $\Delta x \to 0$.

Then $f(x + \Delta x) = A + B \cdot \Delta x \pm h(x, \Delta x) \cdot \Delta x$ and

$$\lim_{x \to 0} f(x + \Delta x) = \lim_{x \to 0} (A + B \cdot \Delta x \pm h(x, \Delta x) \cdot \Delta x)$$
$$= A + B \cdot 0 + 0 = A.$$

Since f is continuous $\lim_{x \to 0} f(x + \Delta x) = f(x)$ and as limits are unique $A = f(x)$. Now substituting in this value for A we have

$$f(x + \Delta x) = f(x) + B \cdot \Delta x \pm h(x, \Delta x) \cdot \Delta x$$

and

$$\frac{f(x + \Delta x) - f(x)}{\Delta x} = \frac{B \cdot \Delta x \pm h(x, \Delta x) \cdot \Delta x}{\Delta x} = B \pm h(x, \Delta x).$$

Hence

$$f'(x) = \lim_{x \to 0} \frac{f(x + \Delta x) - f(x)}{\Delta x} = B \pm \lim_{x \to 0} h(x, \Delta x)$$
$$= B.$$

From this we conclude that $f(x) + f'(x) \cdot \Delta x = A + B \cdot \Delta x$, that $g(x, \Delta x) = \pm h(x, \Delta x)$ and the tangent line is the closest line to the graph of f.

What happens if we use more derivatives? The answer is interesting and you might like to try experimenting to see what happens. We will need such a development later when dealing with curvature. A further interesting point is how naturally the function of **two variables** $g(x, \Delta x)$ entered into our discussion of one variable theory. We only considered the continuity properties of $g(x, \Delta x)$ when we kept the variable x fixed and let the variable Δx tend to 0. From our experience of partial derivatives we might say that g is **partially continuous** (in the second variable). Now that we have the definition of a continuous function of two variables we could investigate what happens if g is a continuous function of both variables. This is another fruitful line of investigation but again we will not pursue it here. Both suggested lines of investigation arose out of rephrasing known results and show how important notation and terminology can be in mathematics. One cannot overvalue the importance of suggestive notation.

For a function of two variables f we see easily that continuity at (x, y) is the same as being able to write f in the following fashion:

$$f(x + \Delta x, y + \Delta y) = f(x, y) + g(x, y, \Delta x, \Delta y)$$

where $g(x, y, \Delta x, \Delta y) \to 0$ as both $\Delta x \to 0$ and $\Delta y \to 0$ and x and y are kept fixed.

To obtain an approximation to $f(x + \Delta x, y + \Delta y)$ when f is assumed to have continuous partial derivatives we recall the **mean value theorem** for functions of one variable. This is normally stated as follows: if f is continuous on $[a, b]$ and differentiable on (a, b), then there exists a point c in (a, b) such that

$$f'(c) = \frac{f(b) - f(a)}{b - a}.$$

If we let $x = a$ and $x + \Delta x = b$ we have that there exists c, $x < c < x + \Delta x$ (assuming $\Delta x > 0$, if $\Delta x < 0$ we are led to the same formula) such that

$$f'(c) = \frac{f(x + \Delta x) - f(x)}{x + \Delta x - x} = \frac{f(x + \Delta x) - f(x)}{\Delta x}.$$

Since $x < c < x + \Delta x$ there exists θ, $0 < \theta < 1$, such that $c = x + \theta \Delta x$ and we have

$$f(x + \Delta x) - f(x) = f'(x + \theta \Delta x) \cdot \Delta x.$$

We now suppose f is defined on an open set of $\mathbf{R}^2$ which contains (x, y) and that f, $\dfrac{\partial f}{\partial x}$ and $\dfrac{\partial f}{\partial y}$ are continuous functions on this open set. Our analysis may look a little complicated but this is only because we are writing out all the details. A second reading shows that at any stage we are using simple results.

Consider

$$f(x + \Delta x, y + \Delta y) - f(x, y).$$

Here we have simultaneously changed x by Δx and y by Δy. We now make this double change in two steps—changing one variable in each step. We have

$$f(x + \Delta x, y + \Delta y) - f(x, y)$$
$$= f(x + \Delta x, y + \Delta y) - f(x, y + \Delta y) + f(x, y + \Delta y) - f(x, y).$$

This trick of changing one variable at a time is often used in the calculus of several variables. We first treat

$$f(x + \Delta x, y + \Delta y) - f(x, y + \Delta y).$$

As in our treatment of partial derivatives we define the function

$$g(x) = f(x, y + \Delta y)$$

(so y and Δy are both to be treated as constants). We have

$$f(x + \Delta x, y + \Delta y) - f(x, y + \Delta y) = g(x + \Delta x) - g(x).$$

Now applying the one variable mean value theorem to g we see that there exists θ, $0 < \theta < 1$, such that

$$g(x + \Delta x) - g(x) = g'(x + \theta \Delta x) \cdot \Delta x.$$

Linear Approximation 49

Now $g'(x) = \dfrac{\partial f}{\partial x}(x, y+\Delta y)$ and by our assumption $\dfrac{\partial f}{\partial x}$ is continuous. Hence there exists another function G such that

$$\frac{\partial f}{\partial x}(x+\theta\Delta x, y+\Delta y) = \frac{\partial f}{\partial x}(x,y) + G(x,y,\Delta x,\Delta y)$$

and for fixed x and y we have $G(x,y,\Delta x,\Delta y) \to 0$ as $\Delta x \to 0$ and $\Delta y \to 0$. Strictly speaking we should suppose at this stage that G also depends on θ since the c which appears in the mean value theorem may not be unique and thus not fully determined by x and Δx. However, we now put all the above information together and arrive at a formula which gives us a suitable approximation and at the same time shows that G does not depend on θ. Our way of writing G was just a little premature. We have

$$\begin{aligned}
f(x+\Delta x, y+\Delta y) - f(x, y+\Delta y) &= g(x+\Delta x) - g(x) \\
&= g'(x+\theta\Delta x) \cdot \Delta x \\
&= \frac{\partial f}{\partial x}(x+\theta\Delta x, y+\Delta y) \cdot \Delta x \\
&= \left(\frac{\partial f}{\partial x}(x,y) + G(x,y,\Delta x,\Delta y)\right)\Delta x.
\end{aligned}$$

We apply exactly the same analysis to $f(x, y+\Delta y) - f(x,y)$ and we find that there exists a function $H(x,y,\Delta y)$ such that

$$f(x, y+\Delta y) - f(x,y) = \left(\frac{\partial f}{\partial y}(x,y) + H(x,y,\Delta y)\right)\Delta y$$

and $H(x,y,\Delta y) \to 0$ as $\Delta y \to 0$. We add these approximations and get

$$\begin{aligned}
&f(x+\Delta x, y+\Delta y) - f(x,y) \\
&= f(x+\Delta x, y+\Delta y) - f(x, y+\Delta y) + f(x, y+\Delta y) - f(x,y) \\
&= \left(\frac{\partial f}{\partial x}(x,y) + G(x,y,\Delta x,\Delta y)\right)\Delta x + \left(\frac{\partial f}{\partial y}(x,y) + H(x,y,\Delta y)\right)\Delta y \\
&= \frac{\partial f}{\partial x}(x,y) \cdot \Delta x + \frac{\partial f}{\partial y}(x,y) \cdot \Delta y + G(x,y,\Delta x,\Delta y) \cdot \Delta x \\
&\quad + H(x,y,\Delta y) \cdot \Delta y.
\end{aligned}$$

We tidy up this formula using the dot product. We have

$$\begin{aligned}
\frac{\partial f}{\partial x}(x,y) \cdot \Delta x + \frac{\partial f}{\partial y}(x,y) \cdot \Delta y &= \left(\frac{\partial f}{\partial x}(x,y), \frac{\partial f}{\partial y}(x,y)\right) \cdot (\Delta x, \Delta y) \\
&= \nabla f(x,y) \cdot (\Delta x, \Delta y).
\end{aligned}$$

Let

$$K(x,y,\Delta x,\Delta y) = (G(x,y,\Delta x,\Delta y), H(x,y,\Delta y)).$$

Then
$$K(x, y, \Delta x, \Delta y) \to (0, 0) \in \mathbf{R}^2 \text{ as } \Delta x \to 0 \text{ and } \Delta y \to 0$$
and $f(x + \Delta x, y + \Delta y)$
$$= f(x, y) + \nabla f(x, y) \cdot (\Delta x, \Delta y) + K(x, y, \Delta x, \Delta y) \cdot (\Delta x, \Delta y).$$
We regard this as the two-dimensional analogue of the formula
$$f(x + \Delta x) = f(x) + f'(x) \cdot \Delta x + g(x, \Delta x) \cdot \Delta x.$$
Our approximation to $f(x + \Delta x, y + \Delta y)$ is
$$f(x, y) + \nabla f(x, y) \cdot (\Delta x, \Delta y)$$
with error $|K(x, y, \Delta x, \Delta y) \cdot (\Delta x, \Delta y)|$.

Example 35. Area = length × breath, $A = lb$. A small increase in l and a small increase in b leads to an approximate increase in A of the size
$$\nabla A \cdot (\Delta l, \Delta b) = \frac{\partial A}{\partial l} \Delta l + \frac{\partial A}{\partial b} \Delta b = b \Delta l + l \Delta b.$$
In this case we can find the actual error (Figure 7.2) since
$$(l + \Delta l)(b + \Delta b) = lb + b \Delta l + l \Delta b + \Delta l \cdot \Delta b.$$

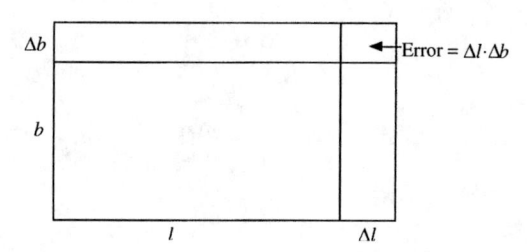

Figure 7.2

Let $l = 12$, $b = 10$, $\Delta l = \Delta b = 1$.
Actual Increase = $(l + \Delta l)(b + \Delta b) - l \cdot b = 13 \cdot 11 - 12 \cdot 10 = 23$.
Approximate increase using partial derivatives
$$= \frac{\partial A}{\partial l}(12, 10) \cdot \Delta b + \frac{\partial A}{\partial b}(12, 10) \cdot \Delta l$$
$$= 10 \cdot 1 + 12 \cdot 1 = 22.$$
Error = $23 - 22 = 1$. Error as a percentage of initial area = $\frac{1}{120} \times 100 = 0.8\%$.

The example just given is rather artificial since we were able to calculate the true value. In practice we usually are not—if we could we would

Linear Approximation

not need an approximation—and this also means that we are not able to find the precise value of the error, and without some idea of the error the approximation is useless. Why? In most cases one has to be satisfied with either an upper bound for the **worst possible error** or the **average error**. It can be shown, if all the second order partial derivatives of f exist and are continuous, that

$$|f(x+\Delta x, y+\Delta y) - f(x,y) - \nabla f(x,y) \cdot (\Delta x, \Delta y)| \leq \frac{1}{2}(|\Delta x| + |\Delta y|)^2 \cdot M$$

where M is the maximum of $\left|\frac{\partial^2 f}{\partial x^2}\right|$, $\left|\frac{\partial^2 f}{\partial x \partial y}\right|$, $\left|\frac{\partial^2 f}{\partial y^2}\right|$ over the rectangle with vertices $(x \pm \Delta x, y \pm \Delta y)$. In practice, there may be a slight problem calculating M. In Example 35, we have $\left|\frac{\partial^2 A}{\partial l^2}\right| = \left|\frac{\partial^2 A}{\partial b^2}\right| = 0$ and $\left|\frac{\partial^2 A}{\partial l \partial b}\right| = 1$, and hence $M = 1$ and Error $\leq \frac{1}{2}(1+1)^2 \cdot 1 = 2$. We conclude that the true value of the increase lies in the interval $[20, 24]$, a statement we can agree with since we have already shown that the true value is 23. Other estimates of the error also follow from Exercise 7.3.

Again as in the one-dimensional case it is easy to see that

$$f(x,y) + \nabla f(x,y) \cdot (\Delta x, \Delta y)$$

is the **unique** approximation with the same quality error, i.e., if

$$f(x+\Delta x, y+\Delta y) = A + B\Delta x + C\Delta y + L(x, y, \Delta x, \Delta y) \cdot (\Delta x, \Delta y)$$

and $L(x, y, \Delta x, \Delta y) \to (0,0) \in \mathbf{R}^2$ as $\Delta x \to 0$ and $\Delta y \to 0$, then $A = f(x,y)$, $B = \frac{\partial f}{\partial x}(x,y)$, $C = \frac{\partial f}{\partial y}(x,y)$ and we obtain the same error.

We now calculate directional derivatives. Let $\vec{v} = (v_1, v_2) \in \mathbf{R}^2$. Let $\Delta x = hv_1$ and $\Delta y = hv_2$. If $h \to 0$, then $\Delta x \to 0$ and $\Delta y \to 0$. Since $(\Delta x, \Delta y) = h\vec{v}$ we have

$$f(x+hv_1, y+hv_2)$$
$$= f((x,y) + h(v_1, v_2))$$
$$= f(x,y) + \frac{\partial f}{\partial x}(x,y) \cdot hv_1 + \frac{\partial f}{\partial y}(x,y) \cdot hv_2 + K(x, y, hv_1, hv_2) \cdot h\vec{v}$$
$$= f(x,y) + h\left(\frac{\partial f}{\partial x}(x,y) \cdot v_1 + \frac{\partial f}{\partial y}(x,y) \cdot v_2\right) + hK(x, y, hv_1, hv_2) \cdot \vec{v}$$

and
$$\lim_{h \to 0} \frac{f((x,y) + h\vec{v}) - f(x,y)}{h}$$
$$= \frac{\partial f}{\partial x}(x,y) \cdot v_1 + \frac{\partial f}{\partial y}(x,y) \cdot v_2 + \lim_{h \to 0} K(x,y,hv_1,hv_2)\vec{v}$$
$$= \frac{\partial f}{\partial x}(x,y) \cdot v_1 + \frac{\partial f}{\partial y}(x,y) \cdot v_2.$$

Hence
$$\frac{\partial f}{\partial \vec{v}} = v_1 \frac{\partial f}{\partial x} + v_2 \frac{\partial f}{\partial y}$$

and since $\frac{\partial f}{\partial x}$ and $\frac{\partial f}{\partial y}$ are both continuous we see that $\frac{\partial f}{\partial \vec{v}}$ is also continuous. We have proved the following result.

Proposition 36. *If f is a continuous function defined on an open subset of $\mathbf{R}^2$ and $\frac{\partial f}{\partial x}$ and $\frac{\partial f}{\partial y}$ both exist and are continuous, then for any vector $\vec{v}$ in $\mathbf{R}^2$ the directional derivative $\frac{\partial f}{\partial \vec{v}}$ exists and is continuous. If $\vec{v} = (v_1, v_2)$ then*
$$\frac{\partial f}{\partial \vec{v}} = v_1 \frac{\partial f}{\partial x} + v_2 \frac{\partial f}{\partial y} = \nabla f \cdot \vec{v}.$$

Exercises

7.1 Let $f(x)$ denote a company's profit expressed as a function of the capital invested x. The ratio
$$\frac{f(x + \Delta x) - f(x)}{\Delta x}$$
when $\Delta x = 1$, is $f(x+1) - f(x)$ and equals the extra or "marginal" profit resulting from the investment of one extra unit of capital given that x units are currently invested. Estimate, using the derivative, the marginal profit when $f(x) = 500x - 2x^2 - 4000$ and $x = 100$. Explain why $f'(x)$ is often called the marginal profit (see Example 43).

7.2 Let $f(x,y) = x^2 y^2 + x^3 + y^4$. By using the value of f and its first order partial derivatives at $(1,1)$ estimate $f(11/10, 9/10)$. Find the error in your approximation and the error as a percentage of $f(1,1)$.

7.3 Using the mean value theorem show, for f a continuously differentiable function of one variable, that
$$\left| f(x + \Delta x) - f(x) - f'(x) \cdot \Delta x \right| \leq |\Delta x| \max_{|\theta| \leq 1} \left| f'(x + \theta \cdot \Delta x) - f'(x) \right|.$$

Linear Approximation

7.4 The total resistance R of two resistors in parallel is given by
$$\frac{1}{R} = \frac{1}{R_1} + \frac{1}{R_2}.$$
Estimate the change in R if R_1 is increased from 20 to 21 ohms and R_2 is decreased from 30 to 29.5 ohms. Find an upper bound for your error.

7.5 If $f : \mathbf{R}^2 \to \mathbf{R}$ is a twice continuously differentiable function and there exists a positive integer n such that $f(tx, ty) = t^n f(x, y)$ for all $(x, y) \in \mathbf{R}^2$ and all $t \in \mathbf{R}$ show that
$$x\frac{\partial f}{\partial x} + y\frac{\partial f}{\partial y} = nf$$
and
$$x^2 \frac{\partial^2 f}{\partial x^2} + 2xy \frac{\partial^2 f}{\partial x \partial y} + y^2 \frac{\partial^2 f}{\partial y^2} = n(n-1)f.$$

7.6 If a, b, c and d are real numbers show that
$$(a^2 + b^2)(c^2 + d^2) - (ac + bd)^2 = (ad - bc)^2$$
Using this result show that
$$|\vec{v} \cdot \vec{w}| \leq \|\vec{v}\| \, \|\vec{w}\|$$
for $\vec{v}$ and $\vec{w}$ vectors in $\mathbf{R}^2$ (this is known as the Cauchy-Schwarz inequality). Hence show that
$$|K(x, y, \Delta x, \Delta y) \cdot (\Delta x, \Delta y)|$$
$$\leq \left(G(x, y, \Delta x, \Delta y)^2 + H(x, y, \Delta y)^2\right)^{1/2} (\Delta x^2 + \Delta y^2)^{1/2}$$
where
$$K(x, y, \Delta x, \Delta y) = (G(x, y, \Delta x, \Delta y), H(x, y, \Delta y)).$$

8

Tangent Lines

Summary. *We consider the direction of maximum increase at a non-critical point and are led to the problem of finding the maximum of a function on a level set. We introduce the Implicit Function Theorem, identify level sets locally with graphs, find the tangent and normal lines to a level set by using a chain rule and compare the rates of change of a function along the tangent line and the level set.*

Until now we have concentrated on the behaviour of a function near a critical point. We now consider functions near a non-critical point. For functions of one variable we recall that $f'(x) > 0$ implies that f is increasing near x and $f'(x) < 0$ implies that f is decreasing near x. Can we find an analogue of this result for functions of two variables?

Let us examine more closely the one-dimensional situation by looking at the graph in Figure 8.1.

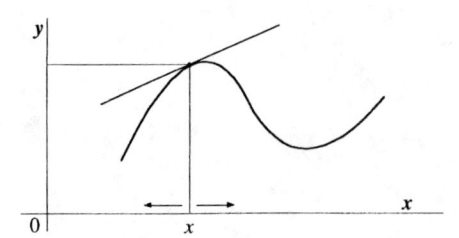

Figure 8.1

At the point x (or indeed at any point on the real line) we can only move in two directions along the x-axis, backwards or forwards. So we can ask in which direction does f increase. If $f'(x) > 0$ then moving forward increases

Tangent Lines

f and if $f'(x) < 0$ then moving backwards increases f. So, at a non-critical point, $f'(x)$ tells us which direction to take in order to increase f.

If we are considering a function of two variables at a non-critical point (x, y) then we have an infinite number, as opposed to two, directions in which to move (Figure 8.2).

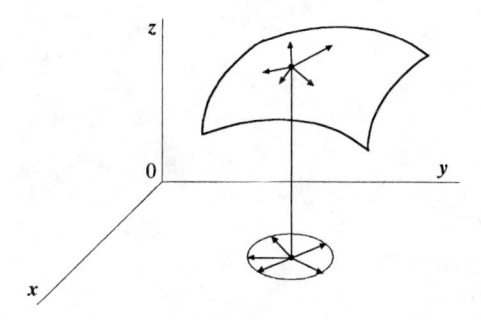

Figure 8.2

Many directions may lead to an increase in f. Can we choose a direction which gives the maximum rate of increase for f? If we look at the unit circle $v_1^2 + v_2^2 = 1$ we see that **each point** gives us a **direction** and we would like to choose a direction $\vec{v} = (v_1, v_2)$ so as to maximize $\dfrac{\partial f}{\partial \vec{v}}(x, y)$. The problem can be posed as follows: find $\vec{v} = (v_1, v_2)$ satisfying $v_1^2 + v_2^2 = 1$ so as to maximize

$$v_1 \frac{\partial f}{\partial x}(x, y) + v_2 \frac{\partial f}{\partial y}(x, y)$$

(where (x, y) is some fixed point with $\nabla f(x, y) \neq (0, 0)$). Having found $\vec{v}$ what is the maximum value of $\dfrac{\partial f}{\partial \vec{v}}(x, y)$? So we obtain a generalization of the one-dimensional result if we can find the maximum of a function on a level set or curve—a problem we have already promised to consider.

To tackle this problem we require further information on the local structure of level sets. A familiar, and as we shall soon see a fairly representable, example of a level set is the graph of a function, i.e., if g is a real-valued function of **one** variable and $f(x, y) = g(x) - y$ then

$$(x, y) \in \text{Graph}(g) \iff y = g(x) \iff (x, y) \in f^{-1}(0).$$

We now state but do not prove the **Implicit Function Theorem**. This tells us that locally most level sets are graphs.

Implicit Function Theorem. *Let $f : U(\text{open}) \subset \mathbf{R}^2 \to \mathbf{R}$ have continuous first order partial derivatives and suppose $\dfrac{\partial f}{\partial y} \neq 0$ at all points of U. If $(a, b) \in U$ and $f(a, b) = c$ then there exists an open interval I containing*

a, a differentiable function $\psi : I \to \mathbf{R}$ satisfying $\psi(a) = b$ and an open set V in $\mathbf{R}^2$ containing (a, b) such that for $x \in I$ and $(x, y) \in V$ we have

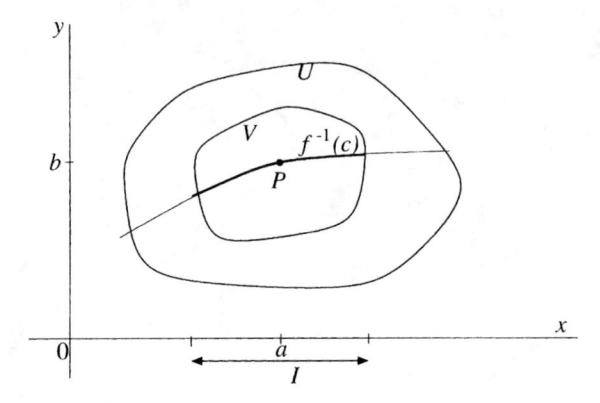

$$f(x, y) = c \iff y = \psi(x) \tag{*}$$

Figure 8.3

The left-hand side of (*) says that (x, y) is on the level set $f^{-1}(c)$ while the right-hand side tells us that (x, y) lies on the graph of ψ. Substituting the right-hand side into the left gives us

$$f(x, \psi(x)) = c$$

for all $x \in I$. Naturally we can exchange the roles of x and y in the Implicit Function Theorem. In this case we require $\dfrac{\partial f}{\partial x} \neq 0$ on U and we obtain a differentiable function $\phi : J \subset \mathbf{R} \to \mathbf{R}$, $\phi(b) = a$, such that $f(\phi(y), y) = c$ for all y in the interval J. Combining these two cases we see that $\nabla f \neq (0, 0)$ on U implies that the level set $f^{-1}(c)$ is locally (i.e., about each point) the graph of a differentiable function of one variable.

The **complete** level set will not always be the graph of a function. Consider $f(x, y) = x^2 + y^2$ and the level set $f^{-1}(1)$. This level set is the unit circle and since vertical and horizontal lines cut the level set more than once it cannot be the graph of a function of either x or y. It may be difficult, and even impossible, to find explicit formulae for ψ and ϕ—this amounts to solving the equation $f(x, y) = c$ for either x or y. When $f(x, y) = x^2 + y^2 = 1$ it is possible and $x = \pm\sqrt{1 - y^2}$ or $y = \pm\sqrt{1 - x^2}$. If P is on the level set and both $\frac{\partial f}{\partial x}(P)$ and $\frac{\partial f}{\partial y}(P)$ are non-zero we can find $y = \psi(x)$ or $x = \phi(y)$ near P (Figure 8.4(c)). If only one of these is non-zero we have no choice (Figures 8.4(a) and 8.4(b)).

In Figure 8.4(a), $y = +\sqrt{1 - x^2} = \psi(x)$, in Figure 8.4(b), $x = +\sqrt{1 - y^2} = \phi(x)$ while in Figure 8.4(c) we can either let $y = -\sqrt{1 - x^2} = \psi(x)$ or $x = -\sqrt{1 - y^2} = \phi(y)$.

Tangent Lines

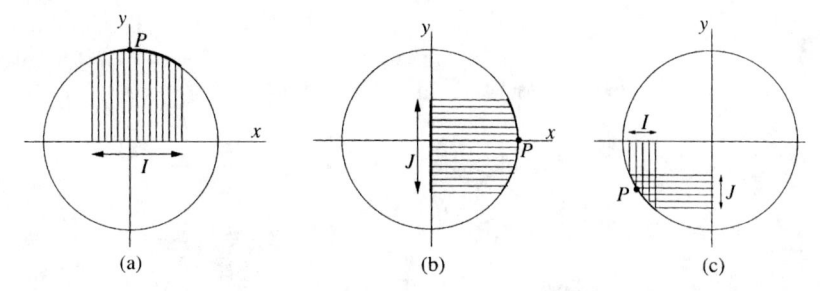

(a)   (b)   (c)

Figure 8.4

From now on we suppose that (∗) in the Implicit Function Theorem holds. The other case can be handled the same way.

We define the **tangent line** to the level set $f^{-1}(c)$ at P to be the tangent line to the graph of ψ at the point P, i.e., the line through $P = (a, b)$ with slope $\psi'(a)$. This line has equation

$$y - b = \psi'(a)(x - a)$$

or

$$y = b + \psi'(a)(x - a). \tag{1}$$

The right-hand side in (1) is just the linear approximation to ψ near a. In one variable calculus this is often presented in an informal manner as "the line that comes closest to the graph near P". This helps develop geometric insight, for example, in seeing that local maxima and minima are critical points.

In two variable calculus we have a further use for the tangent line. Suppose g is a function with continuous first order partial derivatives defined on an open subset of $\mathbf{R}^2$ containing P. The open set will contain $P = (a, b)$ and segments of the tangent line and the graph near P (Figure 8.5(a)) and we consider the restriction of g to both of these sets.

(a)   (b)

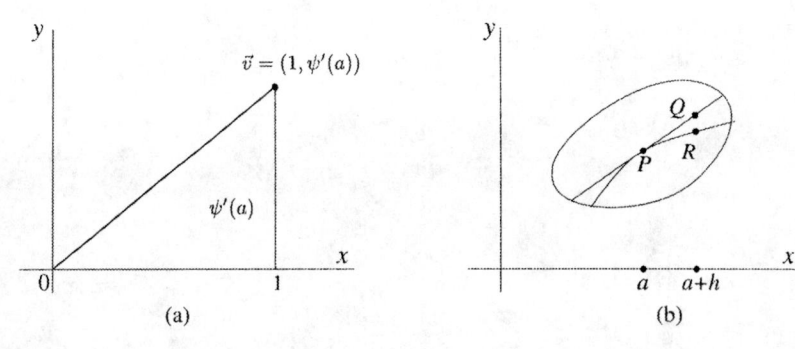

Figure 8.5

We compare the **rates of change** of g along the **tangent line** and along the **level set** $f(x,y) = c$ by taking the values of g above $a+h$, i.e., at Q and R in Figure 8.5(b). Consider the limits of

$$\frac{g(a+h, b+\psi'(a)h) - g(a,b)}{h} \quad \text{and} \quad \frac{g(a+h, \psi(a+h)) - g(a,b)}{h}$$

as $h \to 0$. If $\vec{v} = (1, \psi'(a))$, then

$$\frac{g(a+h, b+\psi'(a)h) - g(a,b)}{h} = \frac{g((a,b) + h(1, \psi'(a))) - g(a,b)}{h}$$

$$= \frac{g(P + h\vec{v}) - g(P)}{h}$$

and the first limit is $\dfrac{\partial g}{\partial \vec{v}}(P)$. Hence the rate of change along the **tangent line** at (a,b) is

$$\frac{\partial g}{\partial \vec{v}}(a,b) = \nabla g(a,b) \cdot \vec{v} = \nabla g(a,b) \cdot (1, \psi'(a)) = \frac{\partial g}{\partial x}(a,b) + \frac{\partial g}{\partial y}(a,b) \cdot \psi'(a) \tag{2}$$

The second limit is $\dfrac{d}{dx}(g(x, \psi(x)))\big|_{x=a}$ and the standard way to differentiate this function uses the **chain rule**. We find the derivative directly using an approach which can be developed to prove the general chain rule. In Chapter 12 we derive this formula intuitively and also give, without proof, the general chain rule.

Now we will find $\dfrac{d}{dx}(g(x, \psi(x)))$. Since ψ is differentiable, we have

$$\psi(x + \Delta x) = \psi(x) + \psi'(x) \cdot \Delta x + k(x, \Delta x) \cdot \Delta x$$

where $k(x, \Delta x) \to 0$ as $\Delta x \to 0$. Hence

$$g(x + \Delta x, \psi(x + \Delta x)) = g(x + \Delta x, \psi(x) + \psi'(x) \cdot \Delta x + k(x, \Delta x) \cdot \Delta x).$$

Since $\dfrac{\partial g}{\partial x}$ and $\dfrac{\partial g}{\partial y}$ are both continuous

$$g(x + \Delta x, y + \Delta y)$$
$$= g(x, y) + \frac{\partial g}{\partial x}(x, y) \cdot \Delta x + \frac{\partial g}{\partial y}(x, y) \cdot \Delta y + l(x, y, \Delta x, \Delta y) \cdot (\Delta x, \Delta y)$$

where

$$l(x, y, \Delta x, \Delta y) \to (0, 0) \in \mathbf{R}^2 \text{ as } \Delta x \to 0 \text{ and } \Delta y \to 0.$$

In this expression for g let $y = \psi(x)$ and $\Delta y = \psi'(x) \cdot \Delta x + k(x, \Delta x) \cdot \Delta x$.

Tangent Lines

Then $\Delta y \to 0$ as $\Delta x \to 0$ and

$$g(x + \Delta x, \psi(x + \Delta x))$$
$$= g(x + \Delta x, \psi(x) + \psi'(x) \cdot \Delta x + k(x, \Delta x) \cdot \Delta x)$$
$$= g(x, \psi(x)) + \frac{\partial g}{\partial x}(x, \psi(x)) \cdot \Delta x$$
$$+ \frac{\partial g}{\partial y}(x, \psi(x)) \cdot (\psi'(x) \cdot \Delta x + k(x, \Delta x) \cdot \Delta x)$$
$$+ l(x, \psi(x), \Delta x, \Delta y) \cdot (\Delta x, \Delta y)$$
$$= g(x, \psi(x)) + \left(\frac{\partial g}{\partial x}(x, \psi(x)) + \frac{\partial g}{\partial y}(x, \psi(x)) \cdot \psi'(x) \right) \Delta x$$
$$+ \frac{\partial g}{\partial y}(x, \psi(x)) \cdot k(x, \Delta x) \cdot \Delta x + l(x, \psi(x), \Delta x, \Delta y) \cdot (\Delta x, \Delta y).$$

Since $k(x, \Delta x) \to 0$ as $\Delta x \to 0$ it follows that

$$\frac{\partial g}{\partial y}(x, \psi(x)) \cdot k(x, \Delta x)$$

and

$$l(x, \psi(x), \Delta x, \Delta y)$$

tend to zero as $\Delta x \to 0$. Let $h(x, \Delta x)$

$$= \frac{\partial g}{\partial y}(x, \psi(x)) \cdot k(x, \Delta x) + l(x, \psi(x), \Delta x, \Delta y) \cdot (1, \psi'(x) + k(x, \Delta x)).$$

The above estimates show that $h(x, \Delta x) \to 0$ as $\Delta x \to 0$. From our expression for $g(x + \Delta x, \psi(x + \Delta x))$ we have

$$g(x + \Delta x, \psi(x + \Delta x))$$
$$= g(x, \psi(x)) + \left(\frac{\partial g}{\partial x}(x, \psi(x)) + \frac{\partial g}{\partial y}(x, \psi(x)) \cdot \psi'(x) \right) \Delta x + h(x, \Delta x) \cdot \Delta x.$$

Since the error term is of the correct type, the **uniqueness** of the approximation implies that

$$\frac{d}{dx}(g(x, \psi(x))) = \frac{\partial g}{\partial x}(x, \psi(x)) + \frac{\partial g}{\partial y}(x, \psi(x)) \cdot \psi'(x). \tag{3}$$

Equations (2) and (3) show that g has the **same rate of change** at P along the level set (or graph) and along the tangent line to the graph. From (1) the tangent line is the line through P in the direction $\vec{v}$. If we let $f = g$ in the above analysis then, by (2) and (3),

$$\frac{\partial f}{\partial \vec{v}}(a, b) = \frac{d}{dx} f(x, \psi(x))\Big|_{x = a}.$$

However, $f(x, \psi(x)) = c$ for all x on the level set and this implies
$$\frac{\partial f}{\partial \vec{v}}(a,b) = \nabla f(a,b) \cdot \vec{v} = 0$$
and $\nabla f(a,b)$ is **perpendicular** to $\vec{v}$ and hence to the **tangent line**. We thus have the following diagram (Figure 8.6).

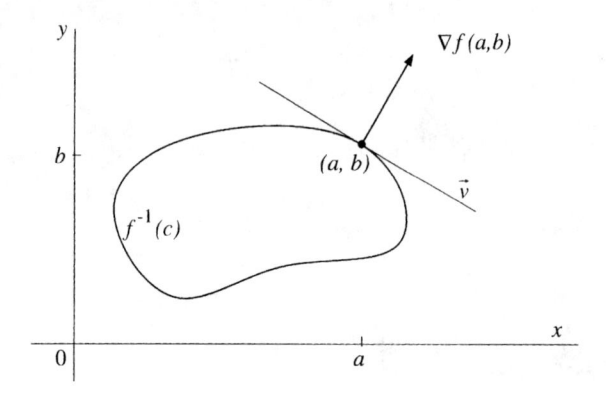

Figure 8.6

The line through P in the direction $\nabla f(a,b)$ is called the **normal line** to the level set. This allows us to find the tangent line to the level set **directly** from the gradient. In particular, we **do not** have to find ψ.

Example 37. Let $f(x,y) = x^2 + 2xy$. We find here the tangent line and the normal line to the level set of f which passes through the point $(1, 2)$. Note that every point in the domain of the function is on some level set of the function. We have $f(1,2) = 1 + 4 = 5$ and $P = (1,2) \in f^{-1}(5)$. Hence

$$\frac{\partial f}{\partial x}(x,y) = 2x + 2y, \qquad \frac{\partial f}{\partial x}(1,2) = 6$$
$$\frac{\partial f}{\partial y}(x,y) = 2x, \qquad \frac{\partial f}{\partial y}(1,2) = 2$$

and $\nabla f(1,2) = (6, 2)$. The normal line to $f^{-1}(5)$ has slope $2/6 = 1/3$ and, since it passes through the point $(1, 2)$, has equation
$$y - 2 = \frac{1}{3}(x - 1) ,$$
i.e., $x - 3y + 5 = 0$. The tangent line has slope -3 and equation
$$y - 2 = -3(x - 1) ,$$
i.e., $3x + y = 5$.

Exercises

8.1 Find $\dfrac{\partial f}{\partial \vec{v}}(x,y)$ where $\vec{v} = (1,3)$ and f is any of the functions in Exercise 2.2.

8.2 Find the equation of the tangent line and the normal line to the level set of the given function f at the indicated point P
 (a) $f(x,y) = 4x^2 + 9y^2$, $P = (-2, -1)$,
 (b) $f(x,y) = 4x^2 - y^2$, $P = (5, -8)$.

8.3 Sketch an open set whose boundary is not the level set of a function with continuous partial derivatives.

8.4 Find all points on the level set $\dfrac{x^2}{4} + \dfrac{y^2}{9} = 1$, near which y (respectively x) can be expressed as a differentiable function of x (respectively y). Find an explicit formula for x (respectively y) at these points.

8.5 If f is a function of two variables with continuous partial derivatives and $g(x) = f(e^x, \cos x)$ find $g'(x)$ in terms of f_x and f_y. If $f(x,y) = \dfrac{x^2}{4} + \dfrac{y^2}{9}$ find $g'(x)$ and $g''(x)$ in terms of x.

8.6 Let $f : U \subset \mathbf{R}^2 \to \mathbf{R}$ and suppose $\nabla f(x,y) \neq (0,0)$ for all $(x,y) \in U$. If $f^{-1}(c) = \operatorname{graph}(\phi) = \operatorname{graph}(\psi)$ show that the tangent lines to both graphs coincide.

9

Method and Examples of Lagrange Multipliers

Summary. *By geometric reasoning we arrive at a method—the method of Lagrange multipliers—of finding the maximum and minimum of a function f on a level set of a function g.*

We now turn to the problem of finding the maximum and minimum of a function on the boundary of an open set. The example of the set $x^2+y^2 \leq 1$ that we gave earlier is in some sense typical. The open set in this case is $x^2 + y^2 < 1$ and its boundary is $x^2 + y^2 = 1$. In this particular case the boundary is the level set of a function and this occurs quite regularly in practice—in fact there is practically no real loss of generality in assuming that this is always the case. So we forget about the open set and just look at the problem of finding the maximum and the minimum of the sufficiently regular function of two variables $f(x, y)$ on a level set of the sufficiently regular function $g(x, y)$. We suppose, in our initial analysis, that the level set has the form $\{(x, y) \in U; g(x, y) = 0\}$ for some open set U in $\mathbf{R}^2$.

The function $f(x, y)$ is also called the **objective function** and the condition $g(x, y) = 0$ a **constraint**—and the problem we are looking at is often called finding the maximum and minimum with constraints. The problem that we considered in earlier chapters is also referred to as the problem of finding maxima and minima without constraints. We will follow the procedure we followed previously and one that is used over and over in mathematics in solving problems: we first experiment by sketching to get the feel of the problem and to get some intuitive idea how to solve it; next we assume as much regularity as we need to get a solution; and finally we go back and justify the assumptions we have made.

Since we have two functions we start by drawing level sets or curves (Figure 9.1). For g we need only use one level set and as we need to know

Method of Lagrange Multipliers

$g(x, y) = 0$

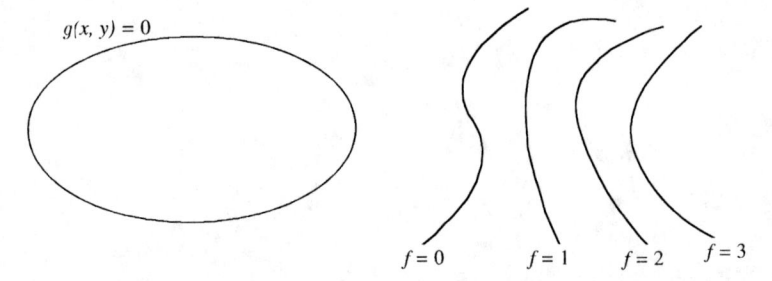

$f = 0$ $f = 1$ $f = 2$ $f = 3$

Figure 9.1

$g(x, y) = 0$

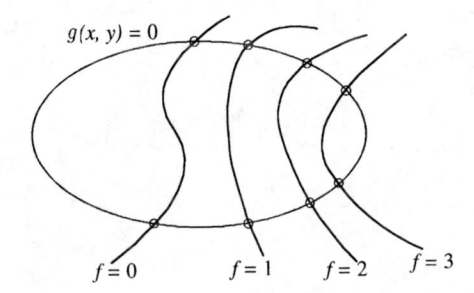

$f = 0$ $f = 1$ $f = 2$ $f = 3$

Figure 9.2

the values of f on $g(x, y) = 0$ we place the level sets of f on top of the level set $g = 0$ (Figure 9.2).

The points on the level set $g(x, y) = 0$ are the only points of interest and we have circled these. From these we see that the maximum of f on $g = 0$ is greater than or equal to 3. We take a closer look at the right-hand side of the sketch and draw more level sets of f near this part (Figure 9.3).

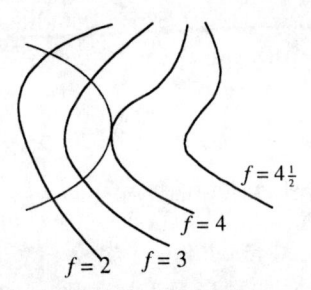

$f = 2$ $f = 3$ $f = 4$ $f = 4\frac{1}{2}$

Figure 9.3

Now $f = 4\frac{1}{2}$ does not intersect $g = 0$ so it cannot be considered and by inspection we see that 4 is a maximum in this part of the level set so it is at least a local maximum. At the point where this is achieved, we have the situation shown in **Figure 9.4**.

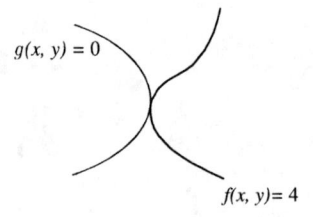

Figure 9.4

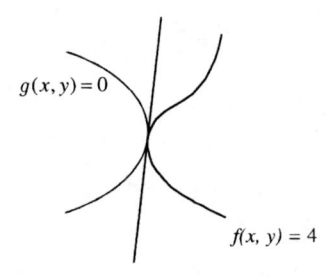

Figure 9.5

If we draw the tangent lines (i.e., the closest lines) to the level sets $g = 0$ and $f = 4$ we see that they coincide (Figure 9.5).

If we draw level sets for other functions f and g we end up in a similar situation and conclude that if f has a local maximum or minimum on the level set $g = 0$ at a point $P = (x_0, y_0)$, then the tangent line to the level set $g = 0$ at P **coincides** with the tangent line to the level set $f - M = 0$ at P where M is the value of f at P (i.e., the local maximum or minimum value). Since we have already identified the tangent line to a level set as the line perpendicular to the gradient at that point we conclude that $\nabla g(x_0, y_0)$ and $\nabla f(x_0, y_0)$ are parallel vectors at a local maximum or minimum (x_0, y_0). If the gradient of g is **non-zero** at (x_0, y_0), then there exists a real number λ such that

$$\nabla f(x_0, y_0) = \lambda \nabla g(x_0, y_0).$$

Our explicit assumption that U is open and the acquired hypothesis that $\nabla g(x, y)$ is non-zero means that the level set behaves like a curve with no end points and the local maximum that we found is similar to finding, in the one variable case, a local maximum at an **internal** point of an interval. However, it is often the case, for instance in Examples 41 and 43, that we wish to find the maximum or minimum of f on the part of $g(x, y) = 0$ which lies in $\overline{U}$ where U is an open subset of $\mathbf{R}^2$ even though g is defined and sufficiently regular on a much larger open set. In such cases, if we still assume that $\nabla g(x, y)$ is non-zero, we may treat the level set as a "curve" whose end points are the points on the level set which lie on the boundary

Method of Lagrange Multipliers 65

of U, i.e., in $\overline{U}\setminus U$. It is also necessary to evaluate the function at these "boundary" points.

The method of finding the maximum and minimum of a function f on a level set, $g(x,y) = 0$, by finding the value of f on the set of all solutions of

$$\nabla f(x,y) = \lambda \nabla g(x,y) \quad \text{and} \quad g(x,y) = 0$$

and comparing these with one another and with the "boundary" values of f is called the **method of Lagrange multipliers**. The real number λ is called a **Lagrange multiplier**.

Example 38. Find the maximum and minimum of $f(x,y) = xy$ on the ellipse $\frac{x^2}{a^2} + \frac{y^2}{b^2} = 1$. Let $g(x,y) = \frac{x^2}{a^2} + \frac{y^2}{b^2} - 1$. We wish to find the maximum and minimum of f on the level set $g(x,y) = 0$. The level set is a closed and bounded subset of $\mathbf{R}^2$ and we thus know **before doing any calculations** that the maximum and minimum of f exist, that they occur among the local maxima and minima of f and hence the method of Lagrange multipliers will yield **at least two** solutions. Now $\nabla f(x,y) = (y,x)$ and $\nabla g(x,y) = \left(\frac{2x}{a^2}, \frac{2y}{b^2}\right) \neq (0,0)$ since $\frac{x^2}{a^2} + \frac{y^2}{b^2} = 1$.

Setting $\nabla f(x,y) = \lambda \nabla g(x,y)$ we have

$$(y,x) = \lambda \left(\frac{2x}{a^2}, \frac{2y}{b^2}\right) = \left(\frac{2\lambda x}{a^2}, \frac{2\lambda y}{b^2}\right)$$

and obtain two equations

$$y = \frac{2\lambda x}{a^2} \quad \text{and} \quad x = \frac{2\lambda y}{b^2}.$$

Do not forget that we also have another equation $\frac{x^2}{a^2} + \frac{y^2}{b^2} = 1$ so that we have three equations in three unknowns, λ, x and y. If any of λ, x or y is zero then all are zero and this contradicts $\frac{x^2}{a^2} + \frac{y^2}{b^2} = 1$. Hence we may divide $y = \frac{2\lambda x}{a^2}$ by $x = \frac{2\lambda y}{b^2}$ to get $\frac{y}{x} = \frac{2\lambda x}{a^2} \cdot \frac{b^2}{2\lambda y} = \frac{xb^2}{ya^2} \Longrightarrow y^2 = x^2\frac{b^2}{a^2}$. Hence $\frac{x^2}{a^2} + \frac{x^2 b^2}{a^2} \cdot \frac{1}{b^2} = 1$ and $x^2 = \frac{a^2}{2}$. So we have solutions $x = \pm\frac{a}{\sqrt{2}}$. Substituting into $y^2 = \frac{x^2 b^2}{a^2}$ gives $y^2 = \frac{x^2 b^2}{a^2} = \frac{b^2}{2}$ and $y = \pm\frac{b}{\sqrt{2}}$. Hence we have four solutions $\left(\frac{a}{\sqrt{2}}, \frac{b}{\sqrt{2}}\right)$, $\left(\frac{a}{\sqrt{2}}, -\frac{b}{\sqrt{2}}\right)$, $\left(-\frac{a}{\sqrt{2}}, \frac{b}{\sqrt{2}}\right)$, and $\left(-\frac{a}{\sqrt{2}}, -\frac{b}{\sqrt{2}}\right)$, as shown in Figure 9.6. By substituting in these values we obtain

$$f\left(\frac{a}{\sqrt{2}}, \frac{b}{\sqrt{2}}\right) = f\left(-\frac{a}{\sqrt{2}}, -\frac{b}{\sqrt{2}}\right) = \frac{ab}{2} \quad \text{maximum}$$

$$f\left(-\frac{a}{\sqrt{2}}, \frac{b}{\sqrt{2}}\right) = f\left(\frac{a}{\sqrt{2}}, -\frac{b}{\sqrt{2}}\right) = -\frac{ab}{2} \quad \text{minimum.}$$

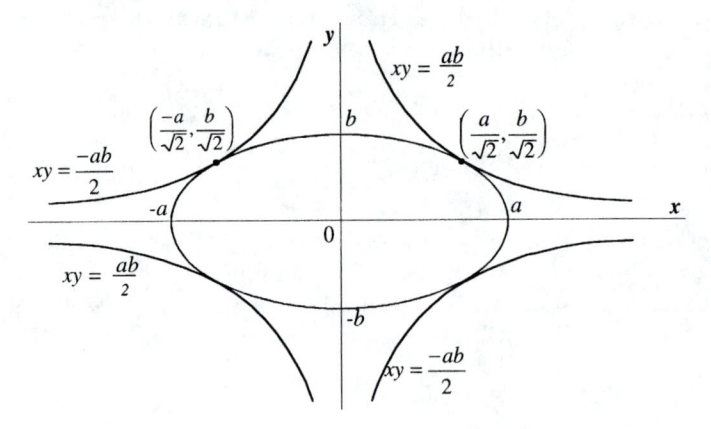

Figure 9.6

Example 39. To find the distance from the origin to the ellipse $x^2 + 2y^2 = 4$. The distance from the point (x, y) to the origin is $\sqrt{x^2 + y^2}$ so we have to find the minimum of $\sqrt{x^2 + y^2}$ when $x^2 + 2y^2 = 4$. Clearly it suffices to find the minimum of $x^2 + y^2$ and to take the square root of the answer. Why?

Let $f(x, y) = x^2 + y^2$ and $g(x, y) = x^2 + 2y^2 - 4$. We apply the method of Lagrange multipliers to find the minimum of $f(x, y)$ on the level set $g(x, y) = 0$. We have

$$\nabla f(x, y) = (2x, 2y) \quad \text{and} \quad \nabla g(x, y) = (2x, 4y) \neq (0, 0)$$

since $x^2 + 2y^2 = 4$. If $\nabla f = \lambda \nabla g$ then

$$2x = 2\lambda x \quad \text{and} \quad 2y = 4\lambda y.$$

Hence $2x - 2\lambda x = 2x(1 - \lambda) = 0$ and $2y - 4\lambda y = 2y(1 - 2\lambda) = 0$. From the first equation we have $x = 0$ or $\lambda = 1$. If $x = 0$ then $x^2 + 2y^2 = 2y^2 = 4$ and $y = \pm\sqrt{2}$ and we have two solutions $(0, \sqrt{2})$ and $(0, -\sqrt{2})$. If $\lambda = 1$ then $2y - 4\lambda y = 2y - 4y = -2y = 0$ and $y = 0$. Hence $x^2 + 2y^2 = x^2 = 4$ and $x = \pm 2$ and we have two more solutions $(2, 0)$ and $(-2, 0)$. From the second equation $2y(1 - 2\lambda) = 0$ we get $y = 0$ or $\lambda = \frac{1}{2}$. We have already found the solutions corresponding to $y = 0$. If $\lambda = \frac{1}{2}$ then $2x - 2\lambda x = 2x - x = 0$

Method of Lagrange Multipliers   67

and $x = 0$ and we have also found these already. Computing f at the four solutions we get $f(0, \sqrt{2}) = f(0, -\sqrt{2}) = 2$ and $f(2, 0) = f(-2, 0) = 4$. So the distance to the ellipse is $\sqrt{2}$. A simple sketch would yield the solution without any calculations.

Example 40. We consider the problem mentioned previously of finding the direction of maximum increase of a function of two variables at a point where the gradient is non-zero. Let P denote the point in question. We have seen that this problem consists of finding the maximum of

$$v_1 \frac{\partial f}{\partial x}(P) + v_2 \frac{\partial f}{\partial y}(P)$$

as $\vec{v} = (v_1, v_2)$ varies over all vectors satisfying $v_1^2 + v_2^2 = 1$. We change our notation to make the constants look like constants and the variables look like variables. We replace

$$\frac{\partial f}{\partial x}(P) \text{ by } a, \quad \frac{\partial f}{\partial y}(P) \text{ by } b, \quad v_1 \text{ by } x \text{ and } v_2 \text{ by } y.$$

Let $h(x, y) = ax + by$ and $g(x, y) = x^2 + y^2 - 1$. We use h to denote the function since we have already used f. We thus have the problem of finding the maximum of h on the set $g = 0$. We have

$$\nabla h(x, y) = (a, b) \quad \text{and} \quad \nabla g(x, y) = (2x, 2y).$$

Since $x^2 + y^2 = 1$, $\nabla g(x, y) \neq (0, 0)$ and for some λ

$$\nabla h(x, y) = \lambda \nabla g(x, y).$$

This gives the two equations $a = 2\lambda x$ and $b = 2\lambda y$ and we also have the equation $x^2 + y^2 = 1$. If $\lambda = 0$ then $a = b = 0$ and hence $\nabla h = \nabla f(P) = (a, b) = (0, 0)$. This is excluded since P is not a critical point of f. If $a \neq 0$ then

$$\frac{b}{a} = \frac{2\lambda y}{2\lambda x} = \frac{y}{x} \text{ and } y = \frac{b}{a}x.$$

Hence

$$x^2 + y^2 = x^2 + \frac{a^2}{b^2}x^2 = x^2\left(1 + \frac{b^2}{a^2}\right) = x^2\left(\frac{a^2 + b^2}{a^2}\right) = 1$$

and

$$x^2 = \frac{a^2}{a^2 + b^2}, \quad x = \pm\frac{a}{\sqrt{a^2 + b^2}}.$$

Now $y = \frac{b}{a}x = \pm\frac{b}{a} \cdot \frac{a}{\sqrt{a^2 + b^2}} = \pm\frac{b}{\sqrt{a^2 + b^2}}$ and so we have two solutions

$$\left(\frac{a}{\sqrt{a^2 + b^2}}, \frac{b}{\sqrt{a^2 + b^2}}\right) \quad \text{and} \quad \left(-\frac{a}{\sqrt{a^2 + b^2}}, -\frac{b}{\sqrt{a^2 + b^2}}\right).$$

Why don't we have four solutions? If we assume $b \neq 0$ we would end up with the same solutions. Now

$$h\left(\frac{a}{\sqrt{a^2+b^2}}, \frac{b}{\sqrt{a^2+b^2}}\right) = \frac{a \cdot a}{\sqrt{a^2+b^2}} + \frac{b \cdot b}{\sqrt{a^2+b^2}}$$
$$= \sqrt{a^2+b^2}$$

and

$$h\left(-\frac{a}{\sqrt{a^2+b^2}}, -\frac{b}{\sqrt{a^2+b^2}}\right) = -\sqrt{a^2+b^2}.$$

Translating this into our original variables we see that the direction for the maximum rate of **increase** of f is

$$\frac{\nabla f(P)}{\|\nabla f(P)\|},$$

i.e., in the direction of $\nabla f(P)$. The maximum rate of increase is

$$\sqrt{a^2+b^2} = \sqrt{\left(\frac{\partial f}{\partial x}(P)\right)^2 + \left(\frac{\partial f}{\partial y}(P)\right)^2} = \|\nabla f(a,b)\|.$$

Similarly the maximum rate of **decrease** is in the direction $-\nabla f(P)$ and the maximum rate of decrease (or the minimum rate of increase) is $-\|\nabla f(P)\|$.

If we consider the level set of f that goes through the point P, i.e.,

$$\{(x,y); f(x,y) = f(P)\}$$

then the normal to this level set at P is $\nabla f(P)$ and hence f has its maximum rate of increase in the direction normal to the level set.

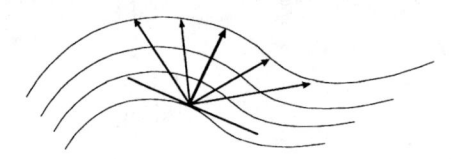

Figure 9.7

In many instances intuitive ideas helped us to begin our mathematical investigations. In this example a formal application of the method of Lagrange multipliers was sufficient and intuitive preliminaries were unnecessary. In such cases an intuitive geometric interpretation of the final result can be useful and instructive.

Any point in the domain is on some level set (of the function) and the direction of maximum increase away from a given point should lead to level sets of **higher value** by the most **direct** (i.e., **shortest**) route. Level sets

Method of Lagrange Multipliers   69

near a non-critical point are often almost "parallel" to one another and, as in Figure 9.7, the shortest route is perpendicular to the tangent line.

Example 41. Find the isosceles triangle of maximum area with fixed perimeter length c (Figure 9.8).

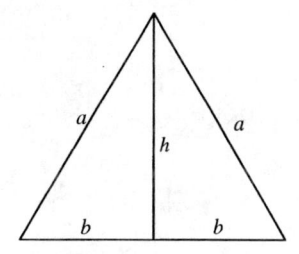

Figure 9.8

We have $c = 2a + 2b$ and we wish to maximize

$$\text{Area} = A(a,b) = bh = b\sqrt{a^2 - b^2}.$$

Let

$$f(a,b) = A^2(a,b) = b^2(a^2 - b^2) = a^2b^2 - b^4$$

and let $g(a,b) = 2a + 2b - c$. Since the shortest path between two points is a straight line we have $2a \geq 2b \geq 0$.

We consider the part of $g(x,y) = 0$ which lies in the open set $a > b > 0$ (Figure 9.9). The boundary points, $(c/2, 0)$ and $(c/4, c/4)$, are obtained by letting $b = 0$ and $a = b$, respectively. At these points we have $A(c/2, 0) = A(c/4, c/4) = 0$.

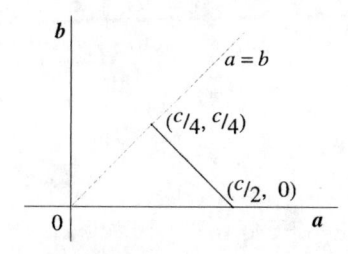

Figure 9.9

Since it is easily seen that $(\max A^2(a,b)) = (\max A(a,b))^2$ our problem reduces to maximizing $f(a,b)$ over all points (a,b) satisfying $g(a,b) = 0$. We have $\nabla g(a,b) = (2,2) \neq (0,0)$ and if $\nabla f(a,b) = \lambda \nabla g(a,b)$ then

$$\nabla f(a,b) = (2ab^2, 2a^2b - 4b^3) = \lambda(2,2) = (2\lambda, 2\lambda).$$

This gives the equations $2ab^2 = 2\lambda$ or $ab^2 = \lambda$ and
$$2a^2b - 4b^3 = 2\lambda \quad \text{or} \quad a^2b - 2b^3 = \lambda.$$
Hence $ab^2 = a^2b - 2b^3$. Since $b \neq 0$ we may divide across by b to get
$$ab = a^2 - 2b^2, \quad \text{i.e.,} \quad a^2 - ab - 2b^2 = 0$$
and
$$a = \frac{b \pm \sqrt{b^2 + 8b^2}}{2} = \frac{b \pm 3b}{2}.$$
Since a and b are both positive this implies $a = \frac{4b}{2} = 2b$, and we obtain an equilateral triangle as the solution.

Exercises

9.1 Use Lagrange multipliers to find the maximum and minimum of the function f subject to the constraint $g = 0$
  (a) $f(x,y) = y^2 - 4xy + 4x^2$, $\quad g(x,y) = x^2 + y^2 - 1$
  (b) $f(x,y) = 2x^2 + xy - y^2 + y$, $\quad g(x,y) = 2x + 3y - 1$
  (c) $f(x,y) = 4x + 3y$, $\quad g(x,y) = x^2 + y^2 - 25$
  (d) $f(x,y) = x^4 + y^4$, $\quad g(x,y) = x^2 + y^2 - 1$.

9.2 Find the maximum and minimum of $3x^2 + 2xy$ inside and on the boundary of the square with vertices $(\pm 1, \pm 1)$.

9.3 If the function $Ax + By$ has its maximum on the set $x^2 + y^2 = a^2$ at (x_0, y_0) show, by drawing level sets, that it has a minimum at the point $(-x_0, -y_0)$.

9.4 Find the maximum and minimum of xy on the set $x^2 + 3y^2 = 6$.

9.5 Find the maximum and minimum of $x^2 - y^2 - 2x + 4y$ on and inside the triangle bounded below by the x-axis, above by the line $y = x + 2$ and on the right by the line $x = 2$.

9.6 Find the dimensions of the rectangles of greatest area that can be inscribed in (a) a semicircle of radius r, (b) a circle of radius r, and (c) the ellipse $\dfrac{x^2}{a^2} + \dfrac{y^2}{b^2} = 1$.

9.7 Use Lagrange multipliers to find the largest possible δ such that
$$\{(x,y); x^2 + (y-1)^2 < \delta^2\} \subset \{(x,y); \frac{x^2}{4} + \frac{y^2}{9} < 1\}$$

Method of Lagrange Multipliers

(see Exercise 1.3).

9.8 (Linear Programming). Sketch the subset of $\mathbf{R}^2$ defined by the inequalities $x \geq 0$, $y \geq 0$, $x + y \leq 4$, $x + 2y \leq 6$. Let $f(x,y) = Ax + By$. By considering different level sets of f show that f achieves its maximum and minimum over E at vertices or corners. Show that f acheives its maximum and minimum at opposite corners of E. Show that the maximum and minimum of f are achieved at unique points of E if and only if the level sets of f are not parallel to any side of the boundary of E. Develop a method so that **one** level set of f leads, by inspection, to the points in E where f achieves its maximum and minimum.

9.9 Use Lagrange multipliers to show
$$(xy)^{1/2} \leq \frac{x+y}{2} \text{ for } x \geq 0, y \geq 0.$$

9.10 Find the maximum and minimum of $x^2 + 2y^2 + 1$ on the set $x^2 + y^2 \leq 1$ without using differential calculus.

9.11 Find the minimum of the function $f(x,y) = 2y$ on the set $3x^2 - y^5 = 0$. Why does the method of Lagrange multipliers not work?

10

Theory and Examples of Lagrange Multipliers

Summary. *Using the implicit function theorem and a special case of the chain rule we provide a mathematical basis for the method of Lagrange multipliers.*

To justify the method of Lagrange multipliers we must show that if $P = (a, b)$ is an "interior" point of the level set $g(x, y) = 0$ at which the function $f(x, y)$ has a local maximum or minimum then there exists a real number λ such that
$$\nabla f(a, b) = \lambda \nabla g(a, b).$$
We first examine a very simple example. Consider the problem of finding the local maxima and minima of $f(x, y) = 2x^2 + y^2$ on the set $x + y = 1$. From the equation $x + y = 1$ we see that $y = 1 - x$ and so the function f takes the values
$$f(x, 1 - x) = 2x^2 + (1 - x)^2 = 3x^2 - 2x + 1$$
on the set $x + y = 1$.

Let $h(x) = f(x, 1 - x) = 3x^2 - 2x + 1$. We have reduced the problem to that of finding the local maxima and minima of a function of one variable. We have
$$\frac{d}{dx}(f(x, 1 - x)) = h'(x) = 6x - 2$$
and
$$\frac{d^2}{dx^2}(f(x, 1 - x)) = h''(x) = 6.$$
Hence h has a critical point at $x = 1/3$. Since $h''(1/3) > 0$ we see that h has a local minimum at $x = 1/3$ and the original function f has a local

Theory of Lagrange Multipliers 73

minimum at $(1/3, 2/3)$. We could also have solved this problem by letting $x = 1 - y$ and analysing the function of one variable $f(1 - y, y)$.

To develop this naïve approach, and naïve approaches often contain the kernel of the final solution, to the general case we should first use the equation $g(x, y) = 0$ to express either y as a function of x or x as a function of y. Substituting this function into f reduces the problem to a one-variable problem. Since we are looking at a **local** maximum or minimum, it is only necessary to consider points on the level set $g(x, y) = 0$ **near** P. So instead of considering $g(x, y) = 0$ as a set of the form shown in Figure 10.1, we consider the set shown in Figure 10.2.

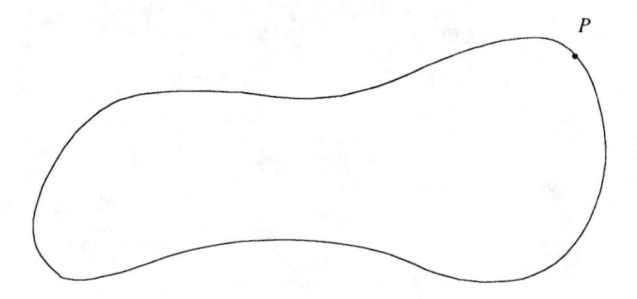

Figure 10.1

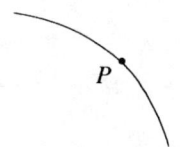

Figure 10.2

This is the situation in which we previously used the implicit function theorem and we do so again here. We suppose $\nabla g \neq (0, 0)$ on the level set and hence near P the level set is a graph. Let $\frac{\partial g}{\partial y}(P) \neq 0$ (the other case is handled in the same way). If $P = (a, b)$ then there exists a differentiable function ψ on an open interval containing a such that $\psi(a) = b$, $g(x, \psi(x)) = 0$ all x near a and the function $x \to f(x, \psi(x))$ has a local maximum or minimum at a. By one-variable calculus $\frac{d}{dx} g(x, \psi(x)) = 0$ all x and $\frac{d}{dx} f(x, \psi(x)) = 0$ when $x = a$.

In Chapter 8 we calculated these derivatives and obtained

$$\frac{d}{dx}\big(g(x, \psi(x))\big) = \frac{\partial g}{\partial x}(x, \psi(x)) + \frac{\partial g}{\partial y}(x, \psi(x)) \cdot \psi'(x)$$

and
$$\frac{d}{dx}\bigl(f(x,\psi(x))\bigr) = \frac{\partial f}{\partial x}(x,\psi(x)) + \frac{\partial f}{\partial y}(x,\psi(x)) \cdot \psi'(x).$$

Letting $x = a$ we get $\psi(x) = b$ and hence
$$\frac{\partial g}{\partial x}(a,b) + \frac{\partial g}{\partial y}(a,b) \cdot \psi'(a) = 0$$

and
$$\frac{\partial f}{\partial x}(a,b) + \frac{\partial f}{\partial y}(a,b) \cdot \psi'(a) = 0.$$

Using the dot product we may rewrite these equations as
$$\nabla g(a,b) \cdot \bigl(1, \psi'(a)\bigr) = 0$$

and
$$\nabla f(a,b) \cdot \bigl(1, \psi'(a)\bigr) = 0.$$

Hence $\nabla g(a,b)$ and $\nabla f(a,b)$ are both perpendicular to the same **non-zero** vector and, since we are in a two-dimensional space, $\nabla f(a,b)$ and $\nabla g(a,b)$ are parallel vectors. Thus there exists a real number λ such that
$$\nabla f(a,b) = \lambda \nabla g(a,b)$$

and we have justified our use of Lagrange multipliers. In particular, we have $\dfrac{\partial f}{\partial y}(a,b) = \lambda \dfrac{\partial f}{\partial y}(a,b)$ and, since $\dfrac{\partial g}{\partial y}(a,b) \neq 0$,
$$\lambda = \frac{\partial f}{\partial y}(a,b) \Big/ \frac{\partial g}{\partial y}(a,b).$$

If $\dfrac{\partial g}{\partial x}(a,b) \neq 0$ we would have obtained
$$\lambda = \frac{\partial f}{\partial x}(a,b) \Big/ \frac{\partial g}{\partial x}(a,b).$$

Our only assumptions on f and g were
(i) f and g both have continuous first order partial derivatives on an open set U containing $g(x,y) = 0$,
(ii) $\nabla g \neq 0$ on the set U.

Now that we have completed our analysis it is appropriate to ask the question—is it worth studying proofs?

Our mathematical analysis here, and in Chapter 8, were both necessary to justify the use of Lagrange multipliers. In the process we obtained new insight into the original problem and unexpected information. We have

— learned that level sets near non-critical points are graphs,
— encountered the implicit function theorem,

Theory of Lagrange Multipliers

— proved a special version of the chain rule (which makes understanding and proving more general versions much easier),

— observed that $\lambda = \dfrac{\partial f}{\partial y}(a,b) \Big/ \dfrac{\partial g}{\partial y}(a,b)$ (from this one can deduce that $\lambda = \dfrac{df}{dg}$ and this sometimes admits a practical interpretation; see Example 43),

— seen that f restricted to the level set $g = 0$ is essentially a function of one real variable and hence nothing similar to saddle points will appear (this is not the case in higher dimensions),

— sufficient information to derive easily a test involving only the second order partial derivatives of f and g in order to identify some, but perhaps not all, local maxima or minima. In fact, all we need to use is the one variable test and the second derivative

$$\frac{d^2}{dx^2} f(x, \phi(x))$$

which we are now able to calculate. Some rather lengthy, but routine, calculations show that (a, b) is a local minimum if

$$\vec{v} H_{(f - \lambda g)(a,b)} {}^t\vec{v} > 0$$

where $\vec{v} = \left(\dfrac{\partial g}{\partial y}(a,b), -\dfrac{\partial g}{\partial x}(a,b) \right)$ is the tangent to the set $g(x,y) = 0$ at (a, b) and λ is the Lagrange multiplier at the point (a, b). In most cases, however, it is not necessary to use this criterion.

I think it is reasonable to say that this proof was worth studying.

Example 42. Find the maximum and minimum of $x^2 + 2y^2 + 1$ on the set $x^2 + y^2 \leq 1$.

This is the type of problem we began this book with, and the solution involves a combination of both methods. On $x^2 + y^2 < 1$ we apply the Hessian method. On $x^2 + y^2 = 1$ we apply Lagrange multipliers and finally we compare all local maxima and minima.

Let $f(x, y) = x^2 + 2y^2 + 1$. We have $\nabla f(x, y) = (2x, 4y)$ and on $x^2 + y^2 < 1$ we have $(0, 0)$ as the only critical point (clearly it has a minimum at this point) but let us check it out anyway. We have

$$H_{f(x,y)} = \begin{pmatrix} 2 & 0 \\ 0 & 4 \end{pmatrix}.$$

Then $\det(H_{f(0,0)}) = 8 > 0$ and since $\dfrac{\partial^2 f}{\partial x^2}(0,0) > 0$ we have a local minimum at $(0, 0)$ and $f(0, 0) = 1$.

Let $g(x,y) = x^2 + y^2 - 1$. We have $\nabla g(x,y) = (2x, 2y)$. If
$$\nabla f(x,y) = \lambda \nabla g(x,y)$$
then
$$2x(1 - \lambda) = 0$$
and
$$2y(1 - 2\lambda) = 0.$$
Hence $x = 0$, or $\lambda = 1$ and $y = 0$ or $\lambda = 1/2$.
If $x = 0$ then $x^2 + y^2 = 1$ implies $y = \pm 1$.
If $\lambda = 1$ then $y = 0$ and $x = \pm 1$.
If $y = 0$ then $x = \pm 1$.
If $\lambda = 1/2$ then $x = 0$.
So the local maxima and minima on $x^2 + y^2 = 1$ can only occur at $(1, 0)$, $(-1, 0)$, $(0, 1)$ or $(0, -1)$.
At these points $f(x, y)$ has the following values:
$$f(1, 0) = f(-1, 0) = 2,$$
$$f(0, 1) = f(0, -1) = 3.$$
Overall 3 is the maximum value, which is achieved at $(0, \pm 1)$; and 1 is the minimum value, which is achieved at $(0, 0)$.

Example 43. Let $f(x,y) = 100 x^{3/4} y^{1/4}$ where $x \geq 0$ and $y \geq 0$. We wish to find the maximum of f subject to the constraint $150x + 250y = 50,000$. In this example we are investigating the function f on the set $\overline{U}$ where $U = \{(x,y) \in \mathbf{R}^2 ; x > 0, y > 0\}$. The boundary of U consists of the positive x-axis and the positive y-axis and these intersect the level set at the points $(1000/3, 0)$ and $(0, 200)$ (see Figure 10.3).

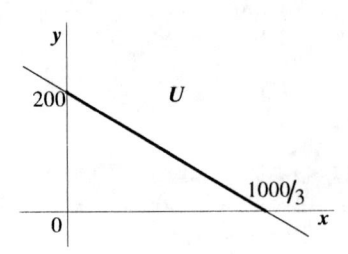

Figure 10.3

Theory of Lagrange Multipliers

We first consider the problem in U and afterwards look at the boundary values. Let $g(x, y) = 150x + 250y - 50,000$. We have $\nabla g(x, y) = (150, 250) \neq (0, 0)$. If $\nabla f(x, y) = \lambda \nabla g(x, y)$ then

$$\nabla f(x, y) = \left(100 \cdot 3/4 \cdot x^{-1/4} y^{1/4}, 100 \cdot 1/4 \cdot x^{3/4} y^{-3/4}\right)$$
$$= \lambda(150, 250) = (150\lambda, 250\lambda).$$

Since x and y are both positive on U we have

$$75 \frac{y^{1/4}}{x^{1/4}} = 150\lambda \quad \text{and} \quad 25 \frac{x^{3/4}}{y^{3/4}} = 250\lambda.$$

Dividing one of these equations by the other gives

$$\frac{75}{25} \frac{y^{1/4}}{x^{1/4}} \frac{y^{3/4}}{x^{3/4}} = \frac{150\lambda}{250\lambda}, \quad \text{i.e.,} \quad \frac{3y}{x} = \frac{3}{5} \quad \text{or} \quad 5y = x.$$

From the equation

$$150x + 250y = 50,000$$

we get

$$150x + \frac{250}{5}x = 50,000$$

$$200x = 50,000 \quad \text{or} \quad x = 250.$$

Hence $y = 50$ and

$$f(250, 50) = 100(250)^{3/4} 50^{1/4} \approx 16,719.$$

Since $f(1000/3, 0) = f(0, 200) = 0$ it follows that $16,719$ is the maximum value of f. In fact, it wasn't really necessary to go through this final check since it is clear that f is zero at every boundary point. In solving the above set of equations we did not need to find λ and this is sometimes the case with Lagrange multiplier type problems. In particular problems, however, λ may have a useful interpretation.

The function $f(x, y) = Cx^\alpha y^{1-\alpha}$, where $0 < \alpha < 1$ and C is a positive constant, is used as a model in economics where it is called the **Cobb-Douglas production function**. We have used the case $\alpha = 3/4$. The variable x represents the number of units of labour and y represents the number of units of capital. In our example each unit of labour costs £150 and each unit of capital costs £250 and the total finance available is £50,000. The function g denotes the amount of finance invested and f represents the level of production. With this investment we have shown that the maximum level of production is 16,719 units. If we return to our equations we see that

$$\lambda = \frac{75}{150} \frac{y^{1/4}}{x^{1/4}} = \frac{1}{2} \frac{(50)^{1/4}}{(250)^{1/4}} \approx 0.334.$$

This value of λ can be interpreted as $\dfrac{df}{dg}$ at $g = 50{,}000$ and represents the **marginal productivity of money**. This means that for every additional £1 spent 0.334 units of the product can be produced. See Exercise 7.1.

When this type of problem is discussed by economists it is common to use notation which recalls or suggests the meaning of the different terms. Thus x, the units of labour, would be denoted by L; y, the units of capital, by K; f, the quantity produced, by Q; and g, the amount of money invested, by M. The problem posed is then to maximize the objective function

$$Q = CL^\alpha K^{1-\alpha}$$

subject to the constraint

$$P_L L + P_K K = M$$

where P_L is the cost of each unit of labour and P_K the cost of each unit of capital. The constant M is the total amount of money available for investment. We have

$$\frac{dQ}{dM} \text{(at the maximum point)} = \lambda.$$

Example 44. As a final application of Lagrange multipliers we derive conditions in terms of a, b and c which imply that the function

$$g(x, y) = ax^2 + 2bxy + cy^2,$$

is positive, negative or neither on the set of all non-zero (x, y) in $\mathbf{R}^2$. We have already solved this problem from first principles in Chapter 4 in order to prove Proposition 24. Our new proof contains a number of simple tricks which may be useful in other situations, and shows how naturally ideas from linear algebra enter into differential calculus and above all shows how different methods can be used to solve the same problem.

If $(x, y) \neq (0, 0)$ let

$$x_1 = \frac{x}{(x^2 + y^2)^{1/2}}, \quad y_1 = \frac{y}{(x^2 + y^2)^{1/2}} \quad \text{and} \quad \alpha = (x^2 + y^2)^{1/2}.$$

Then $\alpha > 0$, $x_1^2 + y_1^2 = 1$ and $(x, y) = \alpha(x_1, y_1) = (\alpha x_1, \alpha y_1)$. Since $g(x, y) = g(\alpha x_1, \alpha y_1) = \alpha^2 g(x_1, y_1)$ it suffices to check the values of g on the set $x_1^2 + y_1^2 = 1$. By the fundamental existence theorem, Chapter 6, g has a maximum and a minimum on $x^2 + y^2 = 1$. Hence

$g(x, y) > 0$ for all $(x, y) \neq (0, 0)$ $\iff$ $\min(g) > 0$ on $x^2 + y^2 = 1$

$g(x, y) < 0$ for all $(x, y) \neq (0, 0)$ $\iff$ $\max(g) < 0$ on $x^2 + y^2 = 1$

g takes positive and negative values $\iff$ $\min(g) < 0 < \max(g)$ on $x^2 + y^2 = 1$.

Theory of Lagrange Multipliers

Our problem is thus reduced to finding the maximum and minimum of $ax^2 + 2bxy + cy^2$ on the set $x^2 + y^2 = 1$. By the method of Lagrange multipliers we see that there exists λ such that

$$(2ax + 2by, 2bx + 2cy) = \lambda(2x, 2y).$$

Hence

$$ax + by = \lambda x \qquad (1)$$
$$bx + cy = \lambda y \qquad (2)$$

First we note that on multiplying Equation (1) by x and Equation (2) by y and adding them together we get

$$ax^2 + 2bxy + cy^2 = \lambda(x^2 + y^2) = \lambda$$

and hence the **values of** λ will give the maximum and minimum of g on $x^2 + y^2 = 1$. From (1) and (2) we get

$$(a - \lambda)x + by = 0$$
$$bx + (c - \lambda)y = 0$$

or in matrix form

$$\begin{pmatrix} a - \lambda & b \\ b & c - \lambda \end{pmatrix} \begin{pmatrix} x \\ y \end{pmatrix} = \begin{pmatrix} 0 \\ 0 \end{pmatrix}. \qquad (3)$$

It is very easy to solve the above equations directly. We prefer, however, to use some linear algebra. The matrix equation (3) has a non-zero solution if and only if

$$\det \begin{pmatrix} a - \lambda & b \\ b & c - \lambda \end{pmatrix} = 0.$$

The values of λ which satisfy this equation are called the **eigenvalues** of the matrix and the non-zero vectors which satisfy (3) are called the associated **eigenvectors**. Since

$$\det \begin{pmatrix} a - \lambda & b \\ b & c - \lambda \end{pmatrix} = \lambda^2 - (a + c)\lambda + ac - b^2$$

there are only two eigenvalues λ_1 and λ_2. Using the fact that

$$\lambda^2 - (\lambda_1 + \lambda_2)\lambda + \lambda_1 \lambda_2 = (\lambda - \lambda_1)(\lambda - \lambda_2)$$

we see that

$$\lambda_1 \lambda_2 = ac - b^2 = \det \begin{pmatrix} a & b \\ b & c \end{pmatrix}$$

and

$$\lambda_1 + \lambda_2 = a + c = \text{trace} \begin{pmatrix} a & b \\ b & c \end{pmatrix}.$$

It is now easy to read off the required properties. Clearly λ_1 and λ_2 have the same sign if and only if $ac - b^2 > 0$ and both are positive (negative) if and only if $a > 0$ ($a < 0$). The pair λ_1 and λ_2 have different signs if and only if $ac - b^2 < 0$. An examination of our proof shows that we have proved a slightly stronger result which we will state only for the positive case. We have shown the following:

*if $g(x, y) > 0$ for all non-zero (x, y) then there exists a **positive** number α such that*

$$g(x, y) \geq \alpha(x^2 + y^2)$$

for all $(x, y) \in \mathbf{R}^2$.

Exercises

10.1 Find the absolute minimum and maximum values of f in the region D:
  (a) $f(x, y) = x^2 + 2xy + 3y^2$, $D = \{(x, y); -2 \leq x \leq 4, -1 \leq y \leq 3\}$
  (b) $f(x, y) = x^2 - 3xy - y^2 + 2y - 6x$, $D = \{(x, y); |x| \leq 3, |y| \leq 2\}$
  (c) $f(x, y) = x^3 + 3xy - y^3$ where D is the triangle with vertices $(1, 2)$, $(1, -2)$ and $(-1, -2)$.

10.2 Solve Exercises 9.1(a) and 9.4 using eigenvalues.

10.3 A flat circular plate has the shape of the ellipse $4x^2 + y^2 \leq 4$. The plate, including the boundary, is heated so that the temperature at any point (x, y) is $T(x, y) = x^2 + 2y^2 - 2x$. Locate the hottest and coldest points on the plate and find the temperature at each of these points.

10.4 A manufacturer's total cost (TC) in producing two items G_1 and G_2 ($G =$ goods) is given by

$$TC = 10Q_1 + 5Q_1Q_2 + 10Q_2$$

where Q_i denotes the quantity of G_i produced. If P_i denotes the price (demand) for one item of G_i the demand functions are given by

$$P_1 = 50 - Q_1 + 3Q_2$$
$$P_2 = 30 + 5Q_1 - Q_2 \ .$$

Explain why the demand functions are somewhat reasonable. Do you think the items are in competition with one another? Does the manufacturer have a monopoly? If the manufacturer has a contract to produce 30 items of either kind, find the number of each item that should be produced in order to maximize profits. Find the marginal profit at this level of production.

11

Tangent Planes

Summary. *We discuss the tangent plane to the graph of a real-valued function of two variables.*

In Chapter 8 we obtained the **tangent line** to the graph of a real-valued function of **one** variable from the linear approximation. In this chapter we use the same approach to define the **tangent plane** to the graph of a real-valued function of two variables.

In the one variable case the linear approximation to $f(x)$ near x_0 is given by
$$y_0 + f'(x_0)(x - x_0)$$
where $y_0 = f(x_0)$ and the tangent line is the straight line with equation
$$y = y_0 + f'(x_0)(x - x_0). \tag{1}$$
If we consider a function f of two variables then the linear approximation to $f(x, y)$ near (x_0, y_0) is
$$z_0 + (x - x_0) \cdot \frac{\partial f}{\partial x}(x_0, y_0) + (y - y_0) \cdot \frac{\partial f}{\partial x}(x_0, y_0)$$
where $z_0 = f(x_0, y_0)$. We define the **tangent plane** to the graph of f at (x_0, y_0, z_0) to be
$$\{(x, y, z); z = z_0 + (x - x_0) \cdot \frac{\partial f}{\partial x}(x_0, y_0) + (y - y_0) \cdot \frac{\partial f}{\partial x}(x_0, y_0)\} \ .$$
In our previous discussion of the linear approximation (Chapter 7) we let $\Delta x = x - x_0$ and $\Delta y = y - y_0$.

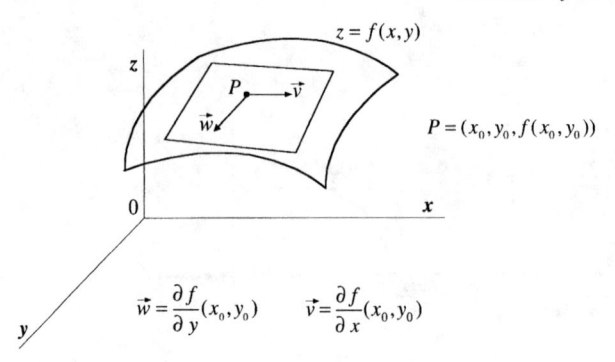

Figure 11.1

Geometrically the tangent plane is the plane through the point $P = (x_0, y_0, z_0)$ which sits closest to the graph of f near P (Figure 11.1).

Cross sections of $\mathbf{R}^3$ through the point P intersect the graph and the tangent plane to produce, in most cases but not always, a curve and a line and it is plausible that this line should be the tangent line to the curve. We consider a few examples. If we take the cross section corresponding to $x = x_0$ we get the curve $y \to f(x_0, y)$ and straight line

$$\{(x_0, y, z); z = z_0 + (y - y_0) \cdot \frac{\partial f}{\partial y}(x_0, y_0)\}.$$

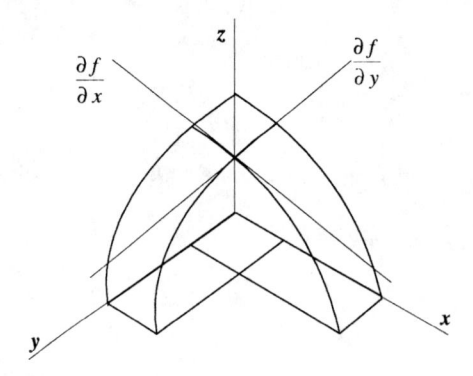

Figure 11.2

From (1) we see that this is the tangent line to the cross section of the graph. If we consider the cross section $y = y_0$ we obtain the curve

$$x \to f(x, y_0)$$

and straight line

$$\{(x, y_0, z); z = z_0 + (x - x_0) \cdot \frac{\partial f}{\partial x}(x_0, y_0)\}$$

Tangent Planes

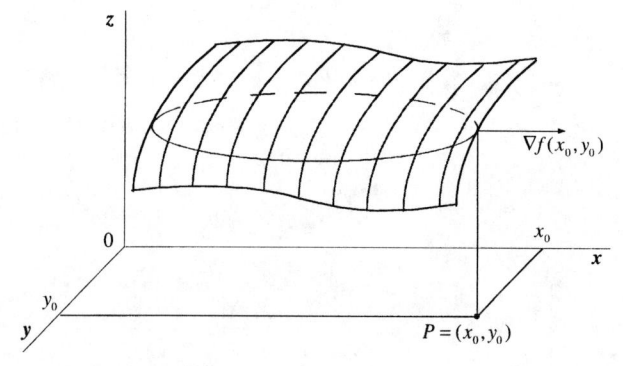

![Figure 11.3 showing a surface with gradient vector ∇f(x₀,y₀) and point P=(x₀,y₀)]

Figure 11.3

and again (1) shows that we obtain a graph and its tangent line. These cross sections are outlined in Figure 11.2.

The final cross section of $\mathbf{R}^3$ that we consider is $z = z_0$. In this case the cross section of the graph is

$$\{(x, y, z_0) \in \mathbf{R}^3; z_0 = f(x, y)\}$$

and if we project this onto $\mathbf{R}^2 \subset \mathbf{R}^3$ (i.e., if we neglect the final coordinate) we get the level set of f.

$$\{(x, y) \in \mathbf{R}^2; z_0 = f(x, y)\}$$

which contains the point (x_0, y_0). The intersection of $z = z_0$ with the tangent plane is the set (Figure 11.3)

$$\{(x, y, z_0) \in \mathbf{R}^3; z_0 = z_0 + (x - x_0) \cdot \frac{\partial f}{\partial x}(x_0, y_0) + (y - y_0) \cdot \frac{\partial f}{\partial x}(x_0, y_0)\}$$
$$= \{(x, y, z_0) \in \mathbf{R}^3; 0 = (x - x_0)\frac{\partial f}{\partial x}(x_0, y_0) + (y - y_0)\frac{\partial f}{\partial x}(x_0, y_0)\}.$$

If we again project onto $\mathbf{R}^2$, by disregarding the final coordinate, and tidy up our notation using the dot product we arrive at

$$\{(x, y) \in \mathbf{R}^2; (x - x_0) \cdot \frac{\partial f}{\partial x}(x_0, y_0) + (y - y_0) \cdot \frac{\partial f}{\partial y}(x_0, y_0) = 0\}$$
$$= \{(x, y) \in \mathbf{R}^2; (x - x_0, y - y_0) \cdot \nabla f(x_0, y_0) = 0\}.$$

If $\nabla f(x_0, y_0) = (0, 0)$ this cross section is all of $\mathbf{R}^2$ but if $\nabla f(x_0, y_0) \neq (0, 0)$ it consists of the straight line through (x_0, y_0) **perpendicular** to the gradient. We have shown in Chapter 8 that this line is the **tangent line** to the **level set** $f(x, y) = z_0$. Thus, in all three cases, we obtain a curve, in fact two graphs and a level set, and the corresponding tangent line.

To define a normal line through a point on the graph we require the concept of dot product in $\mathbf{R}^3$. For vectors (a_1, a_2, a_3) and (b_1, b_2, b_3) in $\mathbf{R}^3$

let
$$(a_1, a_2, a_3) \cdot (b_1, b_2, b_3) = a_1 b_1 + a_2 b_2 + a_3 b_3.$$

All the properties of dot product that we have previously used in $\mathbf{R}^2$ carry over to $\mathbf{R}^3$ in an obvious way. In particular, two vectors are perpendicular in $\mathbf{R}^3$ if and only if their dot product is zero.

Rewriting the equation of the tangent plane through (x_0, y_0, z_0) using the dot product we obtain

$$\{(x, y, z); (x - x_0)\frac{\partial f}{\partial x}(x_0, y_0) + (y - y_0)\frac{\partial f}{\partial x}(x_0, y_0) + (z - z_0)(-1) = 0\}$$
$$= \{(x, y, z); (x - x_0, y - y_0, z - z_0) \cdot (\frac{\partial f}{\partial x}(x_0, y_0), \frac{\partial f}{\partial y}(x_0, y_0), -1) = 0\}.$$

Hence $(\frac{\partial f}{\partial x}(x_0, y_0), \frac{\partial f}{\partial y}(x_0, y_0), -1)$ is perpendicular to the tangent plane and we define the **normal line** to the surface through (x_0, y_0, z_0) as the line in the direction $(\frac{\partial f}{\partial x}(x_0, y_0), \frac{\partial f}{\partial y}(x_0, y_0), -1)$. This line consists of the points

$$\left\{(x_0, y_0, z_0) + t(\frac{\partial f}{\partial x}(x_0, y_0), \frac{\partial f}{\partial x}(x_0, y_0), -1) \; ; \; t \in \mathbf{R}\right\}.$$

Example 45. Let $f(x, y) = x^2 + 2xy$ and $(x_0, y_0) = (1, 2)$. Then

$$\frac{\partial f}{\partial x}(x, y) = 2x + 2y, \qquad \frac{\partial f}{\partial x}(1, 2) = 6$$

$$\frac{\partial f}{\partial x}(x, y) = 2x, \qquad \frac{\partial f}{\partial x}(1, 2) = 2$$

$$z_0 = f(x_0, y_0) = 1 + 2 \cdot 1 \cdot 2 = 5.$$

The tangent plane at $(x_0, y_0, z_0) = (1, 2, 5)$ is

$$z - z_0 = 6(x - x_0) + 2(y - y_0)$$

$$z - 5 = 6x - 6 + 2y - 4 = 6x + 2y - 10$$

$$6x + 2y - z = 5$$

and the normal line is

$$\{(1, 2, 5) + t(6, 2, -1)\} = \{(1 + 6t, 2 + 2t, 5 - t)\}$$

where $t \in \mathbf{R}$ (Figure 11.4).

A unified treatment of tangent lines (to level sets) and tangent planes

Tangent Planes

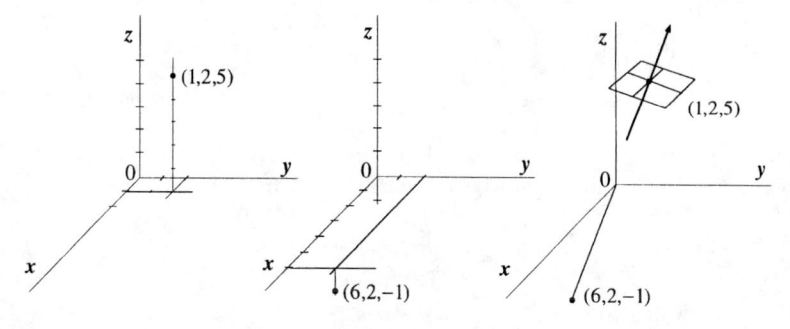

Figure 11.4

(to graphs) is obtained by moving to higher dimensions. If $g : \mathbf{R}^n \to \mathbf{R}$ then the gradient of g, ∇g, is defined as

$$\nabla g = \left(\frac{\partial g}{\partial x_1}, \ldots, \frac{\partial g}{\partial x_n} \right).$$

The **tangent space** to the level set

$$\{x \in \mathbf{R}^n;\ g(x) = c\} = g^{-1}(c)$$

is defined at the point $a \in \mathbf{R}^n$, when $\nabla g(a) \neq 0$, to be

$$\{x \in \mathbf{R}^n;\ (x-a) \cdot \nabla g(a) = 0\}.$$

When $n = 2$ the tangent space is one-dimensional and we use the term **tangent line** and when $n = 3$ the tangent space is two-dimensional and we use the term **tangent plane**. We refer to Chapter 8 for a discussion of tangent lines to level sets.

If $f : \mathbf{R}^n \to \mathbf{R}$ we define $g : \mathbf{R}^{n+1} \to \mathbf{R}$ by letting

$$g(x, y) = f(x) - y \text{ for } x \in \mathbf{R}^n \text{ and } y \in \mathbf{R}.$$

The graph of f coincides with the level set $g^{-1}(0)$, i.e.,

$$\begin{aligned} \operatorname{graph}(f) &= \{(x,y) \in \mathbf{R}^n \times \mathbf{R}\ ;\ f(x) = y\} \\ &= \{(x,y) \in \mathbf{R}^n \times \mathbf{R}\ ;\ g(x,y) = 0\} \\ &= g^{-1}(0) \end{aligned}$$

and we now show that tangent space to the level set coincides with the tangent space to the graph.

When $n = 2$ and $(x_0, y_0) \in \mathbf{R}^2$ then $(x_0, y_0, z_0) \in \operatorname{graph}(f)$ if $z_0 = f(x_0, y_0)$. The tangent plane to $g^{-1}(0)$ at (x_0, y_0, z_0) is the set

$$\{(x,y,z) \in \mathbf{R}^3;\ (x-x_0, y-y_0, z-z_0) \cdot \nabla g(x_0, y_0, z_0) = 0\}.$$

Since $\nabla g(x_0, y_0, z_0) = \left(\dfrac{\partial f}{\partial x}(x_0, y_0), \dfrac{\partial f}{\partial y}(x_0, y_0), -1 \right)$ the tangent plane can be rewritten in the form

$$\{(x, y, z) \in \mathbf{R}^3; (x - x_0) \cdot \frac{\partial f}{\partial x}(x_0, y_0) + (y - y_0) \cdot \frac{\partial f}{\partial y}(x_0, y_0) = z - z_0\}.$$

This is precisely our definition of the tangent plane to the graph of f and also explains the appearance of -1 in the form of the normal line.

Exercises

11.1 Find the equation of the tangent plane and the normal line to the graph of the given function f at the indicated point P
   (a) $f(x, y) = 4x^2 + 9y^2$, $P = (-2, -1, 25)$,
   (b) $f(x, y) = 4x^2 - y^2$, $P = (5, -8, 36)$,
   (c) $f(x, y) = 2e^{-x} \cos y$, $P = (0, \frac{\pi}{3}, 1)$,
   (d) $f(u, v) = \log uv$, $P = (1, 2, \log 2)$.

11.2 Find the point on the paraboloid $z = 4x^2 + 9y^2$ at which the normal line is parallel to the line through $P = (-2, 4, 3)$ and $Q = (5, -1, 2)$.

11.3 Find the equation of the plane in $\mathbf{R}^3$ passing through the points $(1, 2, 3)$ and $(4, 5, 6)$ which is perpendicular to the plane $7x + 8y + 9z = 10$.

11.4 If the function $f : \mathbf{R}^2 \to \mathbf{R}$ has continuous partial derivatives at all points and

$$5x + 4y - 2z = 3$$

is the tangent plane to the graph of f at the point $(1, 1, f(1, 1))$, find $f(1, 1)$, $\frac{\partial f}{\partial x}(1, 1)$ and $\frac{\partial f}{\partial y}(1, 1)$.

11.5 Let $\vec{v}$ and $\vec{w}$ be vectors in $\mathbf{R}^3$ and let ${}^t\vec{w}$ denote the transpose of $\vec{w}$. If $\cdot$ is the dot product and $\circ$ is matrix multiplication, show that $\vec{v} \cdot \vec{w} = \vec{v} \circ {}^t\vec{w}$.

12

The Chain Rule

Summary. *We define differentiable functions from $\mathbf{R}^n \to \mathbf{R}^m$ and present a chain rule which applies in all dimensions. Examples in different dimensions are given.*

A number of chain rules are presented in textbooks. This is due to the variety of notation in use and because different forms arose in different parts of pure and applied mathematics. There is, however, only one chain rule and this applies to all functions in all dimensions.

If A is running **twice** as fast as B and B is running **three times** as fast as C then A is running **six times** as fast as C. This is the chain rule in its simplest form and a good intuitive guide to the fundamental principle underlying every version of the chain rule.

Suppose A, B and C are running in a straight line, having started together at the point 0. Let z, y and x denote the positions of A, B and C, respectively, at time t (Figure 12.1).

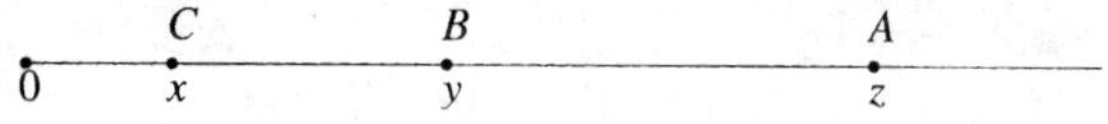

Figure 12.1

Clearly $z = 2y$ and $y = 3x$. Let $f(y) = 2y$ and $g(x) = 3x$. The function f summarises the relationship between A and B, while the function g shows how B depends on C. Hence the composition

$$z = f(y) = f(g(x)) = f(3x) = 2(3x) = 6x$$

describes the relationship between A and C. We have $\dfrac{dz}{dx} = \dfrac{dz}{dy} \cdot \dfrac{dy}{dx} = 2 \cdot 3 = 6$. What happens if A, B and C are not moving at constant speed relative to one another? We maintain the functions f and g to describe the relationships between A and B and B and C, respectively. The composition $f \circ g$ tells us how A changes with respect to C. If f and g are differentiable, then they admit good local linear approximations and may be considered as having **constant speeds**, $f'(y)$ and $g'(x)$, respectively, at times **close** to t.

The constant speed case implies

$$(f \circ g)'(x) = f'(g(x)) \cdot g'(x) = f'(y) \cdot g'(x).$$

Now suppose $f(x,y)$ is a real-valued function of two variables and that x and y are themselves functions of a single variable t, i.e., $x = x(t)$ and $y = y(t)$. This gives the composite function $t \to f(x(t), y(t))$ and the rate of change is the **sum** of the rates of change contributed by each of the coordinates. If $a = x(t_0)$ and $b = y(t_0)$, then the rate of change at t_0 will be

$$\frac{d}{dt}(f(x(t),b))\Big|_{t=t_0} + \frac{d}{dt}(f(a,y(t)))\Big|_{t=t_0}.$$

Applying the one variable chain rule to each of these we obtain

$$\frac{d}{dt}(f(x(t),y(t)))\Big|_{t=t_0} = \frac{\partial f}{\partial x}(a,b) \cdot x'(t_0) + \frac{\partial f}{\partial x}(a,b) \cdot y'(t_0).$$

In a departure from our usual way of proceeding, from the particular to the general, we have found it convenient to go directly to the general case and afterwards to give particular examples.

Since improper use of notation often leads to confusion in applying the chain rule, we proceed by discussing our notation. We will use the following as often as possible. For a real-valued function of two variables, i.e., a function from $\mathbf{R}^2 \to \mathbf{R}$, we use (x,y) to denote the variables in $\mathbf{R}^2$ and we denote the value of the function by $f(x,y)$.

Suppose we have a function from $\mathbf{R}^2 \to \mathbf{R}^2$. In this case we get a pair of real numbers for each value of (x,y) and each of these real numbers will depend on both x and y. So, in reality, we have a pair of **functions** and we may consider our function in the following way:

$$\mathbf{R}^2 \longrightarrow \mathbf{R}^2$$

$$(x,y) \longrightarrow (f(x,y), g(x,y)).$$

Each of these functions may be analysed separately in the same way that we have analysed functions from $\mathbf{R}^2$ into $\mathbf{R}$. Frequently we come across the notation

$$\mathbf{R}^2 \longrightarrow \mathbf{R}^2$$

$$(x,y) \longrightarrow (u,v)$$

The Chain Rule

or

$$(x, y) \longrightarrow (s, t).$$

This means that u is the function in the first coordinate and depends on both x and y so we should write $u(x, y)$ but the variables are often omitted. The same applies to v, s and t.

From now on we prefer to use the variable t for functions of one variable. This will prevent confusion with our use of the variables (x, y) in $\mathbf{R}^2$ and, as we shall see later, it is also helpful, in analysing curves, to think of t as a time variable.

If we have a function from $\mathbf{R}$ into $\mathbf{R}^2$, then the value of the function at any point is a pair of numbers depending on the variable t. So we write it in one of the following ways:

$$t \in \mathbf{R} \longrightarrow \big(f(t), g(t)\big)$$

or

$$t \in \mathbf{R} \longrightarrow \big(x(t), y(t)\big) \qquad (*)$$

or

$$t \in \mathbf{R} \longrightarrow \big(u(t), v(t)\big).$$

The variable t is often omitted and this may lead, if for instance $(*)$ is used, to the confusing statement that (x, y) is a function from $\mathbf{R}$ into $\mathbf{R}^2$.

So far we have been able to get by with partial derivatives, directional derivatives and the gradient and, although we will continue to confine our examples to functions of one and two variables, we have found that a more transparent theory emerges when we introduce a general concept of differentiable function from $\mathbf{R}^n$ into $\mathbf{R}^m$.

If $F : \mathbf{R}^n \to \mathbf{R}^m$ then

$$F(x_1, \ldots, x_n) = \big(f_1(x_1, \ldots, x_n), f_2(x_1, \ldots, x_n), \ldots, f_m(x_1, \ldots, x_n)\big)$$

where each f_i is a real-valued function of n variables. We thus have m independent functions each of which depends on n independent variables. It is reasonable to expect that $m \times n$ real numbers are required to describe the different rates of change at the point $x = (x_1, \ldots, x_n) \in \mathbf{R}^n$. Let $\nabla f_j = \left(\dfrac{\partial f_j}{\partial x_1}, \ldots, \dfrac{\partial f_j}{\partial x_n}\right)$, $1 \leq j \leq m$, denote the gradient of f_j, let $\dfrac{\partial F}{\partial x_i}$,

$D_{x_i}F$ or F_{x_i} denote the ith partial derivative of F for $1 \leq i \leq n$ and let

$$F'(x) = \begin{pmatrix} \nabla f_1 \\ \nabla f_2 \\ \vdots \\ \nabla f_m \end{pmatrix} = \begin{pmatrix} \frac{\partial f_1}{\partial x_1} & \cdots & \frac{\partial f_1}{\partial x_n} \\ \frac{\partial f_2}{\partial x_1} & \cdots & \frac{\partial f_2}{\partial x_n} \\ \vdots & \vdots & \vdots \\ \frac{\partial f_m}{\partial x_1} & \cdots & \frac{\partial f_m}{\partial x_n} \end{pmatrix} = (F_{x_1}, F_{x_2}, \ldots, F_{x_n}).$$

Hence if $F: \mathbf{R}^n \longrightarrow \mathbf{R}^m$, then F' is an $m \times n$ **matrix** (note that the **order is reversed**). Each **row** of F' corresponds to a real-valued function and each **column** to a variable.

Definition 46. *A function $F : \mathbf{R}^n \to \mathbf{R}^m$ is differentiable at $x \in \mathbf{R}^n$ if*

$$F(x + \Delta x) = F(x) + F'(x) \circ {}^t\Delta x + g(x, \Delta x) \circ {}^t\Delta x$$

where $g(x, \Delta x) \to 0$ as $\Delta x \to 0$. If F is differentiable at x we call $F'(x)$ the derivative of F at x.

For $x = (x_1, \ldots, x_n) \in \mathbf{R}^n$, $\Delta x = (\Delta x_1, \ldots, \Delta x_n)$, ${}^t\Delta x$ is the transpose of Δx and the $\circ$ in $F'(x) \circ {}^t\Delta x$ and $g(x, \Delta x) \circ {}^t\Delta x$ is **matrix multiplication** between the $m \times n$ matrices $F'(x)$ and $g(x, \Delta x)$ and the $n \times 1$ matrix ${}^t\Delta x$ and $g(x, \Delta x) \to 0$ as $\Delta x \to 0$ means that each entry in $g(x, \Delta x) \to 0$ as $\Delta x \to 0$.

Definition 46 tells us that a function is differentiable at x if it admits a good linear approximation near x. It is not difficult to show that $F = (f_1, \ldots, f_n)$ is differentiable at x if and only if each f_i is differentiable. A differentiable function has directional derivatives in all directions and, moreover, $\frac{\partial F}{\partial \vec{v}}(x) = F'(x) \circ {}^t\vec{v}$ for all $\vec{v} \in \mathbf{R}^n$. To be comfortable with this general concept of differentiability takes time. We introduced it solely in order to give one clear form of the chain rule but now that we have it we may use it to tidy up conceptually some of the ideas we have been using.

If $f : \mathbf{R} \to \mathbf{R}$ then Definition 46 reduces to the usual definition of differentiable function. If $f : \mathbf{R}^2 \to \mathbf{R}$ is differentiable, then $f'(x, y) = \nabla f(x, y)$. When $f' : \mathbf{R}^2 \to \mathbf{R}^2$ is itself differentiable we write $(f')' = f''$ and, by Definition 46, we have

$$f''(x, y) = \begin{bmatrix} \nabla \left(\frac{\partial f}{\partial x} \right) \\ \nabla \left(\frac{\partial f}{\partial y} \right) \end{bmatrix} = \begin{bmatrix} \frac{\partial^2 f}{\partial x^2} & \frac{\partial^2 f}{\partial y \partial x} \\ \frac{\partial^2 f}{\partial x \partial y} & \frac{\partial^2 f}{\partial y^2} \end{bmatrix} = H_{f(x,y)}.$$

The **Hessian** is the **second derivative** of f.

If $F = (f, g); \mathbf{R}^2 \longrightarrow \mathbf{R}^2$ is such that both f and g are differentiable,

The Chain Rule

then F is differentiable and

$$F'(x,y) = \begin{pmatrix} \nabla f(x,y) \\ \nabla g(x,y) \end{pmatrix} = \begin{pmatrix} \frac{\partial f}{\partial x} & \frac{\partial f}{\partial y} \\ \frac{\partial g}{\partial x} & \frac{\partial g}{\partial y} \end{pmatrix}.$$

A function ϕ from $\mathbf{R}$ into $\mathbf{R}^2$, say

$$\phi : t \longrightarrow (x(t), y(t))$$

is differentiable if $x(t)$ and $y(t)$ are both differentiable and

$$\phi'(t) = \begin{pmatrix} x'(t) \\ y'(t) \end{pmatrix}.$$

We are now in a position to state the **chain rule**.

Proposition 47. *If $F : \mathbf{R}^n \to \mathbf{R}^l$ and $G : \mathbf{R}^l \to \mathbf{R}^m$ are differentiable at $x \in \mathbf{R}^n$ and $y \in \mathbf{R}^l$, respectively, and $y = F(x)$, then*

$$(G \circ F)'(x) = G'(F(x)) \circ F'(x) = G'(y) \circ F'(x)$$

where $\circ$ denotes matrix multiplication.

Some reflection shows that this is quite natural since matrices were developed as a convenient notation for linear mappings and, with this notation, the rule for composition of linear mappings is just **matrix multiplication**. We have

$$\mathbf{R}^n \xrightarrow{F'(x)} \mathbf{R}^l \quad \text{and} \quad \mathbf{R}^l \xrightarrow{G'(y)} \mathbf{R}^m$$

$$\mathbf{R}^n \xrightarrow{G'(y) \circ F'(x)} \mathbf{R}^m.$$

This is the general and only chain rule. The derivative of the composition is the composition of the derivatives. All examples, as we shall now see, are special cases of this rule.

Example 48.

$$\mathbf{R}^2 \xrightarrow{F} \mathbf{R}^2, \quad \mathbf{R}^2 \xrightarrow{G} \mathbf{R}$$

$$(x, y) \longrightarrow (u(x,y), v(x,y)), \quad (u, v) \longrightarrow G(u, v).$$

Consider the mapping

$$G \circ F : \mathbf{R}^2 \longrightarrow \mathbf{R}$$

$$(x, y) \longrightarrow (u(x,y), v(x,y)) \longrightarrow G(u(x,y), v(x,y)).$$

$$(G \circ F)'(x, y) = G'(F(x,y)) \circ F'(x,y).$$

By definition

$$F'(x,y) = \begin{pmatrix} \nabla u(x,y) \\ \nabla v(x,y) \end{pmatrix} = \begin{pmatrix} \frac{\partial u}{\partial x} & \frac{\partial u}{\partial y} \\ \frac{\partial v}{\partial x} & \frac{\partial v}{\partial y} \end{pmatrix}$$

$$G'(u,v) = \nabla G = \left(\frac{\partial G}{\partial u}, \frac{\partial G}{\partial v} \right)$$

so

$$(G \circ F)'(x,y) = \left(\frac{\partial G}{\partial u}, \frac{\partial G}{\partial v} \right) \begin{pmatrix} \frac{\partial u}{\partial x} & \frac{\partial u}{\partial y} \\ \frac{\partial v}{\partial x} & \frac{\partial v}{\partial y} \end{pmatrix}$$

$$= \left(\frac{\partial G}{\partial u} \cdot \frac{\partial u}{\partial x} + \frac{\partial G}{\partial v} \frac{\partial v}{\partial x} \;,\; \frac{\partial G}{\partial u} \cdot \frac{\partial u}{\partial y} + \frac{\partial G}{\partial v} \cdot \frac{\partial v}{\partial y} \right).$$

If we let $H = G \circ F$, then $H' = \left(\frac{\partial H}{\partial x}, \frac{\partial H}{\partial y} \right) = (G \circ F)'$ so, comparing coordinates, we get

$$\frac{\partial H}{\partial x} = \frac{\partial G}{\partial u} \cdot \frac{\partial u}{\partial x} + \frac{\partial G}{\partial v} \cdot \frac{\partial v}{\partial x}$$

and

$$\frac{\partial H}{\partial y} = \frac{\partial G}{\partial u} \cdot \frac{\partial u}{\partial y} + \frac{\partial G}{\partial v} \cdot \frac{\partial v}{\partial y}.$$

The variables u and v appear at an intermediate stage and when one writes out H fully it is a function of x and y—the variables u and v do not appear. For this reason they are often called **silent variables**. In the expression for $\frac{\partial H}{\partial x}$ they appear to "cancel", i.e., they appear above and below the same number of times. To remember the above formula one fills it in, in the following way:

$$\frac{\partial H}{\partial x} = \frac{\partial G}{\partial x} + \frac{\partial G}{\partial x}$$

(one term for each of the silent variables—start with ∂G, where G is the final function in the expression for H, and end with ∂x in each term).
Then fill in the silent variables

$$\frac{\quad}{\quad} = \frac{\partial u}{\partial u} + \frac{\partial v}{\partial v}.$$

Putting them together we get

$$\frac{\partial H}{\partial x} = \frac{\partial G}{\partial u} \cdot \frac{\partial u}{\partial x} + \frac{\partial G}{\partial v} \cdot \frac{\partial v}{\partial y}.$$

The Chain Rule

The final expression is obtained by writing the right-hand side in terms of x and y.

Example 49.

$$\mathbf{R}^2 \xrightarrow{F} \mathbf{R}^2, \quad \mathbf{R}^2 \xrightarrow{G} \mathbf{R}^2$$

$$(x, y) \longrightarrow (x^2 + y^2, x^2 - y^2),$$
$$\parallel$$
$$(u, v) \longrightarrow (e^{uv}, e^{-uv})$$

$$(G \circ F)(x, y) = G(x^2 + y^2, x^2 - y^2) = (e^{x^4 - y^4}, e^{y^4 - x^4})$$

$$(G \circ F)'(x, y) = \begin{pmatrix} \nabla(e^{x^4-y^4}) \\ \nabla(e^{y^4-x^4}) \end{pmatrix} = \begin{pmatrix} 4x^3 e^{x^4-y^4} & -4y^3 e^{x^4-y^4} \\ -4x^3 e^{y^4-x^4} & 4y^3 e^{y^4-x^4} \end{pmatrix}$$

$$F'(x, y) = \begin{pmatrix} \nabla(x^2 + y^2) \\ \nabla(x^2 - y^2) \end{pmatrix} = \begin{pmatrix} 2x & 2y \\ 2x & -2y \end{pmatrix}$$

$$G'(u, v) = \begin{pmatrix} \nabla(e^{uv}) \\ \nabla(e^{-uv}) \end{pmatrix} = \begin{pmatrix} ve^{uv} & ue^{uv} \\ -ve^{-uv} & -ue^{-uv} \end{pmatrix}$$

$$(G \circ F)'(x, y) = G'((F(x, y)) \circ F'(x, y) = G'(x^2 + y^2, x^2 - y^2) \circ F'(x, y)$$

$$= \begin{pmatrix} ve^{uv} & ue^{uv} \\ -ve^{-uv} & -ue^{-uv} \end{pmatrix} \begin{pmatrix} 2x & 2y \\ 2x & -2y \end{pmatrix}$$

$$= \begin{pmatrix} (x^2 - y^2)e^{x^4-y^4} & (x^2 + y^2)e^{x^4-y^4} \\ (y^2 - x^2)e^{y^4-x^4} & -(x^2 + y^2)e^{y^4-x^4} \end{pmatrix} \begin{pmatrix} 2x & 2y \\ 2x & -2y \end{pmatrix}$$

$$= \begin{pmatrix} 4x^3 e^{x^4-y^4} & -4y^3 e^{x^4-y^4} \\ -4x^3 e^{y^4-x^4} & 4y^3 e^{y^4-x^4} \end{pmatrix}.$$

We have found the derivative in two different ways and verified the chain rule.

Example 50. Let $T(x, y)$ denote the temperature at the point (x, y) on the flat circular disc of radius 1. Suppose the rates of change of T along horizontal and vertical lines are given by

$$\frac{\partial T}{\partial x} = 2x - y \quad \text{and} \quad \frac{\partial T}{\partial y} = 2y - x.$$

We wish to find the rate of change of T along a circle of radius r, $0 < r < 1$. A typical point on the circle of radius r has coordinates $(r\cos\theta, r\sin\theta)$ and so we want to find $\frac{d}{d\theta} T(r\cos\theta, r\sin\theta)$. We have the decomposition

$$\mathbf{R} \xrightarrow{F} \mathbf{R}^2, \quad \mathbf{R}^2 \xrightarrow{T} \mathbf{R}$$

$$\theta \longrightarrow (r\cos\theta, r\sin\theta)$$
$$\parallel$$
$$(x, y) \longrightarrow T(x, y)$$

where we let $F(\theta) = (r\cos\theta, r\sin\theta)$. Hence

$$\frac{d}{d\theta} T(r\cos\theta, r\sin\theta) = (T \circ F)'(\theta) = T'(F(\theta)) \circ F'(\theta)$$

$$= \left(\frac{\partial T}{\partial x}(F(\theta)), \frac{\partial T}{\partial y}(F(\theta)) \right) \begin{pmatrix} -r\sin\theta \\ r\cos\theta \end{pmatrix}$$

$$= (2r\cos\theta - r\sin\theta) \cdot (-r\sin\theta) + (2r\sin\theta - r\cos\theta) \cdot r\cos\theta$$

$$= r^2(\sin^2\theta - \cos^2\theta).$$

In discussing Lagrange multipliers, we worked out the derivative of a function ϕ satisfying $f(x, \phi(x)) = 0$ and $\frac{\partial f}{\partial y} \neq 0$. This function can be written as the composition of two functions

$$x \longrightarrow (x, \phi(x)) \quad \text{and} \quad (x, y) \longrightarrow f(x, y)$$

$$\mathbf{R} \longrightarrow \mathbf{R}^2, \quad \mathbf{R}^2 \longrightarrow \mathbf{R}.$$

Let us call the first function g. Then

$$f \circ g(x) = f(x, \phi(x)) = 0.$$

By the chain rule $(f \circ g)'(x) = f'(g(x)) \circ g'(x) = 0$. We have

$$f'(x, y) = \nabla f(x, y) = \left(\frac{\partial f}{\partial x}(x, y), \frac{\partial f}{\partial y}(x, y) \right),$$

$$\nabla f(x, \phi(x)) = \left(\frac{\partial f}{\partial x}(x, \phi(x)), \frac{\partial f}{\partial y}(x, \phi(x)) \right),$$

$$g'(x) = \begin{pmatrix} 1 \\ \phi'(x) \end{pmatrix}.$$

Hence

$$(f \circ g)'(x) = 0 = \left(\frac{\partial f}{\partial x}(x, \phi(x)), \frac{\partial f}{\partial y}(x, \phi(x)) \right) \begin{pmatrix} 1 \\ \phi'(x) \end{pmatrix}$$

$$= \frac{\partial f}{\partial x}(x, \phi(x)) + \phi'(x) \frac{\partial f}{\partial y}(x, \phi(x)).$$

We assumed $\frac{\partial f}{\partial y}(x, \phi(x)) \neq 0$ and so

$$\phi'(x) = \frac{-\frac{\partial f}{\partial x}(x, \phi(x))}{\frac{\partial f}{\partial y}(x, \phi(x))}.$$

We have often looked to the one variable theory for guidance in order to proceed with the several variables case. This example goes in the other direction. What we have just been doing is often called **implicit differentiation** in the one variable theory (see Exercise 12.1) and the several variables theory, including the use of the chain rule, explains what is behind this method of differentiating. The adjective "implicit" is used here and in the implicit function theorem for precisely the same reason.

The chain rule is rather elementary if one follows some rather simple rules. First write down the domain and range of each function involved and then label each function. This first step, however, can cause difficulties due to the careless fashion in which these problems are often posed. Consider the following problem: let $w = 2xy$, $x = s^2 + t^2$, $y = s/t$ find $\frac{\partial w}{\partial s}$? To calculate w one needs to know x and y, and to know x and y one first needs s and t. Each of these describes a function and we must **start** with s and t. We have

$$w : (s,t) \in \mathbf{R}^2 \xrightarrow{f} (s^2 + t^2, s/t) \in \mathbf{R}^2$$
$$\parallel$$
$$(x,y) \in \mathbf{R}^2 \xrightarrow{g} 2xy \in \mathbf{R}.$$

We have labelled these functions f and g. Hence $w = g \circ f$ and

$$w'(s,t) = \left[\frac{\partial w}{\partial s}, \frac{\partial w}{\partial t}\right] = g'(f(s,t)) \circ f'(x,y)$$

$$= [2y, 2x] \begin{bmatrix} 2s & 2t \\ 1/t & -s/t^2 \end{bmatrix}$$

$$= [\frac{2s}{t}, 2s^2 + 2t^2] \begin{bmatrix} 2s & 2t \\ 1/t & -s/t^2 \end{bmatrix}$$

$$= [6s^2/t + 2t, -s^2/t^2 + 2s]$$

and $\dfrac{\partial w}{\partial s} = \dfrac{6s^2}{t} + 2t$.

We have already applied the chain rule in our study of Lagrange multipliers. Later we use it to obtain a **change of variables** rule in integration theory.

Exercises

12.1 Let $f(x,y) = x^3 + y^3$. If ϕ is a differentiable function of x satisfying $f(x, \phi(x)) = 1$ find $\phi'(2^{-1/3})$ and $\phi''(2^{-1/3})$ directly (i.e., by solving $x^3 + y^3 = 1$ and differentiating), by using partial derivatives, and by "implicit differentiation".

12.2 Let $w = x^2y - y^2$, $x = \sin t$, $y = e^t$. Write out the domain and range of each function. Find $\dfrac{dw}{dt}$ directly and then use the chain rule to verify your result.

12.3 Let $f(x,y) = (x^2 + y^2, xy)$ and $g(s,t) = \bigl(\sin(st), \cos(st)\bigr)$. Find all first order partial derivatives of $f \circ g$ and $g \circ f$.

12.4 Let $f(x,y)$ be a function of two variables and let

$$g(x) = f(\sin x, e^x + \cos x).$$

Find $g'(x)$ in terms of f_x, f_y and the derivatives of $\sin x$ and $e^x + \cos x$ by (a) using the chain rule and (b) using approximations.

12.5 Given that $\dfrac{\partial f}{\partial x}(1,0) = 2$ and $\dfrac{\partial f}{\partial y}(1,0) = 3$ find the rate of change of $f(x,y)$ along the parabola $y = x^2 + x - 2$ at the point $(1,0)$.

12.6 If the function $f(x,y)$ satisfies the equation

$$x^2 \frac{\partial^2 f}{\partial x^2} + y^2 \frac{\partial^2 f}{\partial y^2} + x\frac{\partial f}{\partial x} + y\frac{\partial f}{\partial y} = 0$$

show that $f(e^s, e^t)$ satisfies
$$\frac{\partial^2 f}{\partial s^2} + \frac{\partial^2 f}{\partial t^2} = 0.$$

12.7 If f and g are twice differentiable functions of a single variable and
$$w = f(s - ct) + g(s + ct)$$
show that
$$\frac{\partial^2 w}{\partial t^2} = c^2 \frac{\partial^2 w}{\partial s^2}.$$

12.8 The voltage in a circuit that satisfies Ohm's law, $V = IR$, is decreasing as the battery wears out and the resistance is increasing as the temperature increases. From the equation
$$\frac{dV}{dt} = \frac{\partial V}{\partial I} \cdot \frac{dI}{dt} + \frac{\partial V}{\partial R} \cdot \frac{dR}{dt}$$
find the rate of change in the current, $\frac{dI}{dt}$, when $R = 500$, $I = 0.03$, $\frac{dR}{dt} = 0.6$ and $\frac{dV}{dt} = -0.01$.

12.9 Let (a, b) and (c, d) denote open intervals in $\mathbf{R}$. Show that
$$U := (a, b) \times (c, d) = \{(x, y) \in \mathbf{R}^2; x \in (a, b), y \in (c, d)\}$$
is an open subset of $\mathbf{R}^2$. Let $M(U) =$
$$\{f : U \to \mathbf{R}; \text{second order partial derivatives exist and } \frac{\partial^2 f}{\partial x \partial y} = \frac{\partial^2 f}{\partial y \partial x}\}.$$

Show that $M(U)$ has the following properties:
(i) if f and g are in $M(U)$ then $f + g$, $f - g$ and $f \cdot g$ are also in $M(U)$, and if $g \neq 0$, then $f/g \in M(U)$,
(ii) if $h(x, y) = x$ and $k(x, y) = y$ for all $(x, y) \in U$ then h and k belong to $M(U)$,
(iii) if $f \in M(U)$ and $\phi : \mathbf{R} \to \mathbf{R}$ is a twice continuously differentiable function then $\phi \circ f \in M(U)$ (use the chain rule).
By using Exercise 6.1 and the above prove that
$$f(x, y) = \sin^2\left(\frac{xy}{x^2 + y^2 + 1}\right)$$
belongs to $M(U)$. Verify your result by calculating the second order mixed partial derivatives of f.

13

Directed Curves

Summary. *We define directed curves, parametrized curves, unit tangents and normals, obtain a formula for the length of a curve and show that any directed curve admits a unit speed parametrization.*

We have come across curves in our investigations in many different settings, e.g., as the graph of a function of one variable, as the boundary of an open set and as the level set of a function of two variables. We used the term loosely and as an aid to our intuition but in each case we could, if pressed, fall back on a mathematical definition. We provide, in the next few chapters, a systematic study of these objects by considering mappings from $\mathbf{R}$ to $\mathbf{R}^2$. A new feature that we could previously afford to ignore is introduced here. This is the idea of **direction** along a curve. This concept is fundamental in integration theory in higher dimensions and while it can appear initially as an unnecessary complication it is a good idea to master this concept by studying elementary curves in simple settings. Keeping track of directions and checking that parametrizations give the correct sense of direction from the very beginning leads to a better appreciation of the overall picture.

A curve may be defined as the path of a continuously moving point or particle. This gives us a perfectly adequate mathematical definition and a good intuitive guide. For the purposes we have in mind more regularity is required and, instead of adding extra conditions as we need them, we build them into our definition. Our definition applies to a large collection of curves and the reader who follows our approach should have little difficulty in developing a more general theory which applies to curves excluded by our definition, e.g., curves with corners, piecewise differentiable curves and closed curves oriented clockwise. Our definition of **directed curve** may appear rather technical, but the inclusion of technical conditions in the definition leads to the rapid development of a smooth theory.

Directed Curves

We denote by t the variable in $\mathbf{R}$ and think of it as a **time** variable. We consider only functions defined on a closed interval $[a, b]$, an interval of time, and consider $P(t)$, the image of t by P, as the **position** of a **particle** in $\mathbf{R}^2$ at time t.

Definition 51. *A **directed curve** in $\mathbf{R}^2$ is a triple $\{\Gamma, A, B\}$ consisting of a set of points Γ (sometimes called a **geometric curve**) in $\mathbf{R}^2$ and points A and B in Γ, called respectively the **initial** and the **final points** of Γ, for which there exists a mapping $P : [a, b] \to \mathbf{R}^2$ satisfying the following conditions:*

(i) $P([a, b]) = \Gamma$,
(ii) $P(a) = A$ and $P(b) = B$,
(iii) P' and P'' both exist and are continuous on $[a, b]$,
(iv) $P'(t) \neq 0$ for all $t \in [a, b]$,
(v) P is one to one (injective) on $[a, b)$ and $(a, b]$,
(vi) if $A = B$ then, as t increases, $P(t)$ moves around Γ in an **anticlockwise** or **counterclockwise** direction.

We call P a **parametrization** of the directed curve, $\{\Gamma, A, B\}$.

A mapping $P : [a, b] \to \mathbf{R}^2$ which satisfies (iii), (iv), (v) and (vi) with $P(a)$ and $P(b)$ replacing A and B is called a **parametrized curve**.

Thus each parametrized curve $P : [a, b] \to \mathbf{R}^2$ specifies precisely one directed curve $\{P([a, b]), P(a), P(b)\}$ and P is a parametrization of this directed curve. We briefly discuss the conditions in Definition 51. Condition (iii) is merely a regularity condition on P similar to those encountered in max/min problems. Note, however, that the derivatives are defined at a and b and we are thus assuming that P can be extended to an open interval containing $[a, b]$ as a twice differentiable function. Condition (iv) guarantees the existence of a tangent at each point of the curve Γ but excludes curves with corners. Condition (v) tells us that the curve does not cross itself or cover itself more than once. Our rather strange way of writing this condition was necessary to include the case $P(a) = A = B = P(b)$. A curve with this property is called a **closed curve** (Figure 13.1(b)). Condition (i) says that the image of P is a geometric curve while condition (ii), when $A \neq B$, says that the **direction** is from A to B. If $A = B$, then condition (vi) says that the closed curve is directed in an **anticlockwise direction**. This condition simplifies our analysis and avoids ambiguities. Of course, closed curves directed in a clockwise direction do occur in practice and when confronted can be dealt with by making fairly obvious adjustments. But for us, **all** closed curves are directed in an anticlockwise fashion. If $A \neq B$, then $\{\Gamma, B, A\}$ is also a directed curve which occupies the same space as $\{\Gamma, A, B\}$ but the sense of direction has been reversed (Figure 13.1(a)).

If $\{\Gamma, A, A\}$ is a closed curve and $B \in \Gamma$, then $\{\Gamma, B, B\}$ is also a closed curve and we may treat $\{\Gamma, A, A\}$ and $\{\Gamma, B, B\}$ as identical objects.

Functions of Two Variables

(a)

(b)

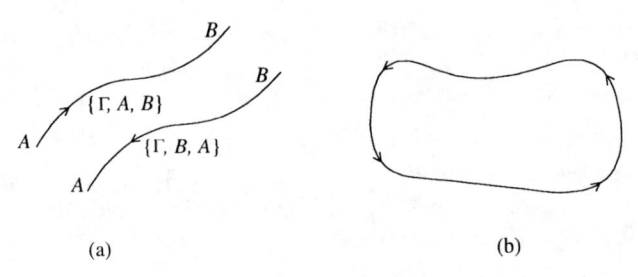

Figure 13.1

Since the notation $\{\Gamma, A, B\}$ is rather cumbersome we agree to write Γ in place of $\{\Gamma, A, B\}$ from now on and to use the term **curve** when either the initial and final points are clearly specified or when the comments and results apply simultaneously to a directed curve and to the curve obtained by reversing the direction.

A directed curve Γ admits many different parametrizations. If P is a parametrization of the directed curve Γ, then $P'(t)$ is called the **velocity** at time t. Since $P'(t) \neq 0$

$$P'(t) = \|P'(t)\| \cdot \frac{P'(t)}{\|P'(t)\|}$$

and $\dfrac{P'(t)}{\|P'(t)\|}$ is a **unit vector** in the direction of motion—i.e., a **unit tangent** to the directed curve at $P(t)$—and is denoted by $T(t)$.

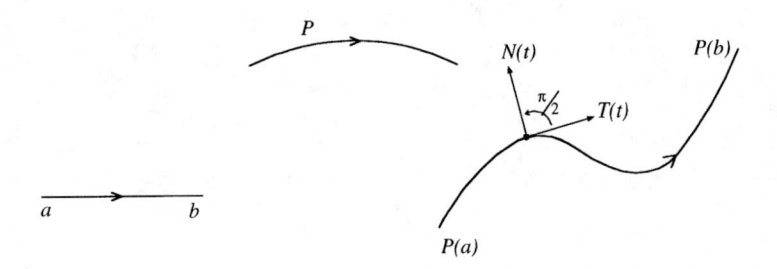

Figure 13.2

Of the two unit vectors perpendicular to $T(t)$ we call the one obtained by rotating $T(t)$ through an angle $+\pi/2$ (i.e., by rotating the tangent vector in an **anticlockwise** direction) the **unit normal** at $P(t)$ and denote it by $N(t)$ (Figure 13.2). If $P(t) = \big(x(t), y(t)\big)$ then it is easily checked that

$$N(t) = \frac{\big(-y'(t), x'(t)\big)}{\|P'(t)\|}.$$

The line through $P(t)$ in the direction $T(t)$ is called the **tangent line** to

Directed Curves

Γ at $P(t)$ and the line through $P(t)$ perpendicular to the tangent line is called the **normal line** to Γ at $P(t)$. We call $\|P'(t)\|$ the **speed** since it tells us how fast the particle is travelling. Hence we have

$$\textbf{Velocity} = \textbf{Speed} \times \textbf{Unit Tangent}.$$

The condition $P'(t) \neq 0$ for all t only excludes curves with corners. To go around a corner one must slow down so that at the corner—and we mean a sharp corner—the speed is zero, otherwise you would not stay on the curve.

The curve in Figure 13.3 has a corner at the origin. The mapping $P(t) = (t^3, |t|^3)$, $-1 \leq t \leq 1$, is differentiable but since $P'(0) = (0,0)$, it is not a parametrization.

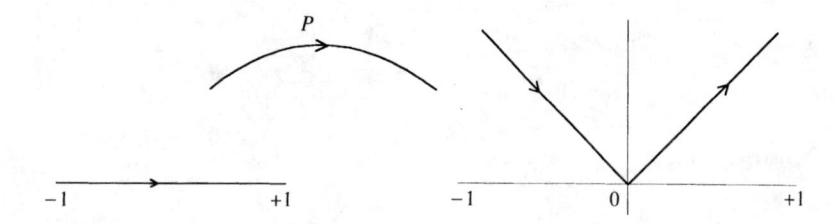

Figure 13.3

In terms of the coordinate functions $x(t)$ and $y(t)$ we have

$$\|P'(t)\| = \|(x'(t), y'(t))\| = \sqrt{x'(t)^2 + y'(t)^2}$$

and

$$T(t) = \frac{P'(t)}{\|P'(t)\|} = \frac{(x'(t), y'(t))}{\sqrt{x'(t)^2 + y'(t)^2}}.$$

We first find the length of the curve $\Gamma = P([a, b])$. We have (Figure 13.4)

$$(\Delta l)^2 \approx (\Delta x)^2 + (\Delta y)^2 = \left[\left(\frac{\Delta x}{\Delta t}\right)^2 + \left(\frac{\Delta y}{\Delta t}\right)^2\right](\Delta t)^2$$

where we have used the symbol $\approx$ to denote approximately equal.

$$(\Delta l)^2 \approx (\Delta x)^2 + (\Delta y)^2$$

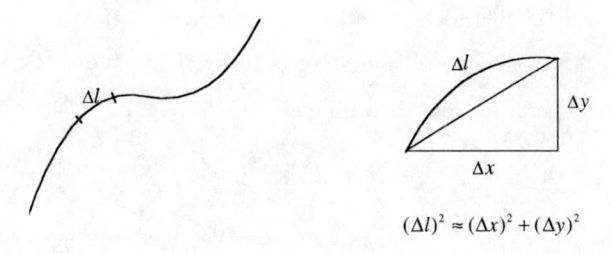

Figure 13.4

So
$$l(\Gamma) = \text{length }(\Gamma) \approx \sum |\Delta l|$$
$$\approx \sum \left[\left(\frac{\Delta x}{\Delta t}\right)^2 + \left(\frac{\Delta y}{\Delta t}\right)^2\right]^{1/2} (\Delta t)$$
$$\to \int_a^b (x'(t)^2 + y'(t)^2)^{1/2}\, dt.$$

Hence
$$l(\Gamma) = \int_a^b \sqrt{x'(t)^2 + y'(t)^2}\, dt = \int_a^b \|P'(t)\|\, dt.$$

This is not unexpected since
$$\|P'(t)\|\Delta t = \text{speed} \times \text{time} = \text{distance}$$
and the distance travelled by the particle is the length of the curve.

Example 52. Let $P(t) = (\cos t + t\sin t, \sin t - t\cos t)$, $t \in [0,1]$. Then
$$P'(t) = (-\sin t + \sin t + t\cos t, \cos t - \cos t + t\sin t)$$
$$= (t\cos t, t\sin t)$$
and $\|P'(t)\| = (t^2 \cos^2 t + t^2 \sin^2 t)^{1/2} = (t^2)^{1/2} = t$.

Hence
$$l(P[0,1]) = \int_0^1 \|P'(t)\|\, dt = \int_0^1 t\, dt = \left.\frac{t^2}{2}\right]_0^1 = \frac{1}{2}.$$

It is often useful to have a parametrization with $\|P'(t)\| = 1$ for all t. For obvious reasons we call such a parametrization a **unit speed parametrization** and we now show how, starting with any parametrization P, we can construct a unit speed parametrization.

We begin by defining a distance function along the directed curve. This function is traditionally denoted by s (Figure 13.5). Let

$s(t) = $ distance the particle has travelled up to time t
$= $ length of the curve between $P(a)$ and $P(t)$.

By our formula for length we have
$$s(t) = \int_a^t \|P'(x)\|\, dx.$$

(Note that we cannot use t as the variable in the integral since we are already using it as a limit of integration).

Directed Curves

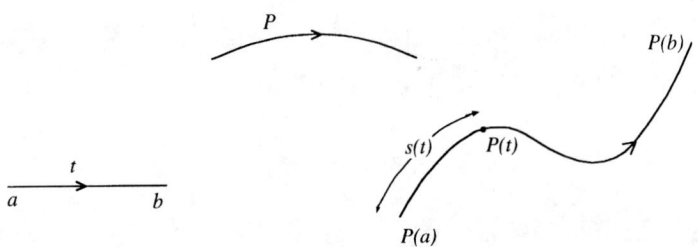

Figure 13.5

To study the function s we require the one variable version of the **fundamental theorem of calculus**: if f is a continuous function on $[a, b]$ and $F(t) = \int_a^t f(x)\, dx$ then F is differentiable and $F'(t) = f(t)$.

If we apply this to the function s we have

$$s'(t) = \|P'(t)\| \quad \text{for all } t \in [a, b].$$

We also know $s(a) = 0$, since the distance travelled in 0 units of time is zero, and $s(b) =$ length of Γ and we denote this by l. Since $s'(t) = \|P'(t)\| > 0$ it follows that s is **strictly increasing** and so gives a **bijective** (i.e., a one to one and onto) mapping from $[a, b]$ onto $[0, l]$. The inverse mapping s^{-1} is a differentiable function from $[0, l]$ onto $[a, b]$ (Figure 13.6). By considering the composition of these mappings we get a unit speed parametrization of our directed curve.

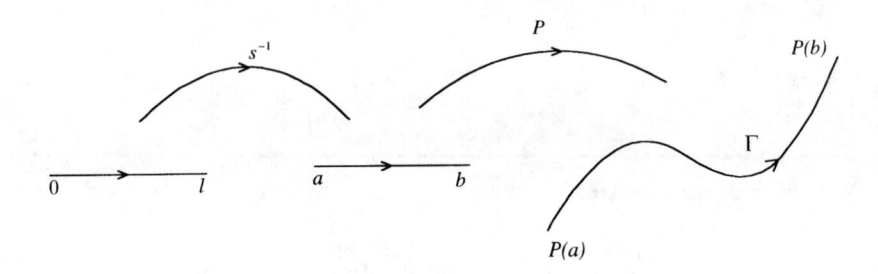

Figure 13.6

To show that this is actually the case we must show that

$$\left\|\frac{d}{dt}(P \circ s^{-1})(t)\right\| = 1 \quad \text{for all } t \in [0, l].$$

By the chain rule we have, since $s \circ s^{-1}(t) = t$,

$$\frac{d}{dt}(s \circ s^{-1})(t) = \frac{d}{dt}(t) = 1 = s'(s^{-1}(t)) \cdot (s^{-1})'(t).$$

Hence
$$(s^{-1})'(t) = \frac{1}{s'(s^{-1}(t))}$$
$$= \frac{1}{\|P'(s^{-1}(t))\|}$$
$$\frac{d}{dt}(P \circ s^{-1})(t) = P'(s^{-1}(t)) \cdot (s^{-1})'(t) = \frac{P'(s^{-1}(t))}{\|P'(s^{-1}(t))\|}$$

and $\|(P \circ s^{-1})'(t)\| = 1$ for all $t \in [0, l]$, i.e., $P \circ s^{-1}$ is a unit speed parametrization of the original curve.

Exercises

13.1 Find the equations of the unit tangent and the unit normal to the curve defined on $[-\pi/3, \pi/3]$ by $P(t) = (\tan t, 1/\cos t)$ at $t = \pi/6$.

13.2 Find the length of the curve parametrized by
$$P(t) = (\cos^3 t, \sin^3 t), \quad 0 \le t \le \pi/8.$$
Can you draw this curve? Is it a unit speed parametrization?

13.3 Find the length of the curve
$$P(t) = (e^t \cos t, e^t \sin t), \quad 0 \le t \le 1.$$

13.4 Show that
$$\det \begin{pmatrix} T(t) \\ N(t) \end{pmatrix} > 0$$
for any parametrized curve.

13.5 Obtain unit speed parametrizations of the curves defined by
(a) $t \longrightarrow (e^t \cos t, e^t \sin t)$, $t \in [0, \frac{\pi}{2}]$,
(b) $t \longrightarrow (\cos 2t, \sin 2t)$, $t \in [0, \frac{\pi}{2}]$,
(c) $t \longrightarrow (t, \cosh t)$, $t \in [0, 1]$.

13.6 Let $P : [-1, +1] \to \mathbf{R}^2$ denote a parametrization of the curve $\{\Gamma, A, B\}$. Let N, T and l denote the normal, tangent and length of $\{\Gamma, A, B\}$. Let $Q(t) = P(-t)$. Show that Q is a parametrization of $\{\Gamma, B, A\}$ if and only if $A \ne B$. If $A \ne B$ and N_1, T_1 and l_1 denote the normal, tangent and length of $\{\Gamma, B, A\}$, respectively, show that $N_1 = -N$, $T_1 = -T$ and $l = l_1$.

13.7 If $P : [a,b] \to \mathbf{R}^2$ is a parametrized curve show that $P([a,b])$ is a closed subset of $\mathbf{R}^2$. (It is thus necessary not to confuse the concepts of closed curve and closed set.)

If $f : P([a,b]) \to \mathbf{R}$ is a continuous function show, using Theorem 1, that f has a maximum and a minimum on $P([a,b])$.

14

Curvature

Summary. *We define the curvature at a point on a unit speed parametrized curve as the rate of change of the angle between the positive x-axis and the tangent. We obtain explicit formulae for the curvature of an arbitrarily directed curve and give some examples.*

Euclidean geometry is based on the fundamental plane curves—the straight line and the circle. Within this limited setting many important concepts, which we still use today, were discovered and developed. For instance, in this book we have used tangents, perpendicular vectors, Pythagoras' theorem and other ideas which were first considered by the ancient Greeks many many years before coordinate systems and the differential calculus were discovered.

We have used the term *limited* since circles and straight lines are rather predictable objects. A straight line is fully determined by any two distinct points and a circle by three distinct points. Moreover, the intersection of circles and straight lines is again described by a small finite number of points. This had to be the case since, without the means to deal with functions which enjoy a smooth but unpredictable variation, i.e., without the differential calculus, only rather regular functions and curves could be analysed. The advent of the differential calculus led to the possibility of studying a vastly increased collection of curves with a wide variety of new methods. This area of mathematics is called differential geometry.

We take a more geometric point of view in this chapter by assigning to each point on a directed curve a real number which we call the **curvature**. The absolute value of this number turns out to be the reciprocal of the radius of the circle through the point which sits closest to the curve. This geometric approach may also be considered as a natural development of the

Curvature

problem of finding the closest straight line (the tangent line) to the curve. Since the first derivative is the slope of the tangent line, we should not be surprised if the second derivative appears in the formula for curvature. Any concept of curvature should intuitively assign curvature 0 to all points on a straight line and the curvature should be constant at all points on a circle. If a circle has large radius, then the curvature should be small—indeed it is often useful to think of a straight line as a circle with infinite radius. We shall use parametrizations to study curvature but clearly our definition of curvature should not depend on the particular parametrization used. We start with a **unit speed parametrization**

$$P : [a, b] \to \Gamma \subset \mathbf{R}^2.$$

Since $\|P'(t)\| = 1$ we have $P'(t) = T(t)$, the unit tangent. Suppose the positive x-axis makes an angle $\theta(t)$ with the vector $P'(t)$ (Figure 14.1).

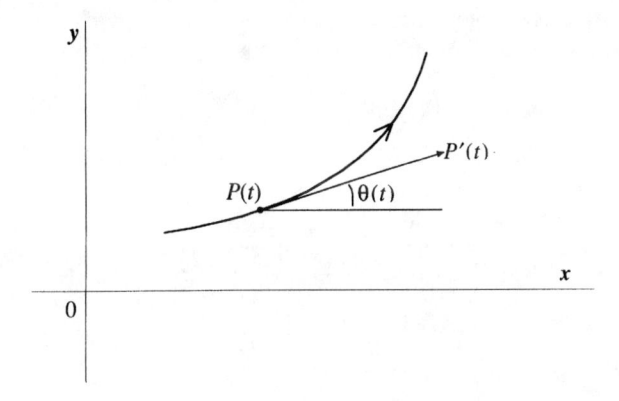

Figure 14.1

The rate of change of θ as we move along the curve, at unit speed, is our measure of curvature.

Definition 53. $k(t)$, the **curvature** at the point $P(t)$, is defined as $\dfrac{d\theta}{dt}$ evaluated at t. We call $|k(t)|$ the **absolute curvature** at $P(t)$.

For example if Γ is a circle of radius r, then the length of the circumference is $2\pi r$ and so a particle will, at unit speed, travel over the full circle in $2\pi r$ units of time. Since we can place this circle at any place in $\mathbf{R}^2$ we may suppose that the centre is at the origin and so the rate of change of θ is constant. Since θ changes from 0 to 2π it follows that the absolute curvature is

$$\frac{2\pi}{2\pi r} = \frac{1}{r} \qquad \left(\frac{\text{change in angle}}{\text{time}} = \text{curvature} \right).$$

We have $P'(t) = \big(x'(t), y'(t)\big)$ and if $x'(t) \neq 0$ then

$$\tan\big(\theta(t)\big) = \frac{y'(t)}{x'(t)} \quad \text{and} \quad \theta(t) = \tan^{-1}\left(\frac{y'(t)}{x'(t)}\right).$$

Hence

$$\theta'(t) = \frac{1}{1 + \left(\dfrac{y'(t)}{x'(t)}\right)^2} \cdot \frac{y''(t)x'(t) - x''(t)y'(t)}{x'(t)^2}$$

$$= \frac{y''(t)x'(t) - x''(t)y'(t)}{x'(t)^2 + y'(t)^2}.$$

Since P is a unit speed parametrization

$$\|P'(t)\|^2 = 1 = x'(t)^2 + y'(t)^2.$$

When $y'(t) \neq 0$ we obtain the same answer. Hence if $P(t) = \big(x(t), y(t)\big)$ is a **unit speed parametrization** of a directed curve, then

$$k(t) = y''(t)x'(t) - x''(t)y'(t).$$

We now consider the case where P is an **arbitrary parametrization**. Since $P \circ s^{-1}$ is a unit speed parametrization we may use the formula we have just established. Now $P \circ s^{-1} = (x \circ s^{-1}, y \circ s^{-1})$ and applying the chain rule and the fact that $(s^{-1})'(t) = \dfrac{1}{s'\big(s^{-1}(t)\big)}$ we get

$$(x \circ s^{-1})'(t) = \frac{x'\big(s^{-1}(t)\big)}{s'\big(s^{-1}(t)\big)} = \frac{x'}{s'}\big(s^{-1}(t)\big).$$

Repeating this formula implies

$$(x \circ s^{-1})''(t) = \frac{x''s' - s''x'}{(s')^3}\big(s^{-1}(t)\big).$$

Using a similar formula for $y \circ s^{-1}$ we find

$$k(t) = \frac{(y''s' - s''y')x' - (x''s' - s''x')y'}{(s')^4}\big(s^{-1}(t)\big)$$

$$= \frac{(y''x' - x''y')}{(s')^3}\big(s^{-1}(t)\big).$$

This gives the curvature at $P\big(s^{-1}(t)\big)$ where t is an arbitrary point in $[0, l]$. Hence, if we consider the curvature at an arbitrary point on $[a, b]$, the domain of P, and use the fact that

$$s'(t) = \big(x'(t)^2 + y'(t)^2\big)^{1/2}$$

Curvature

we obtain the curvature at a point $P(t)$, $t \in [a, b]$,

$$\kappa = \frac{y''x' - x''y'}{\left|(x')^2 + (y')^2\right|^{3/2}}$$

where all derivatives are evaluated at t.

This formula is very practical since it can be computed directly from **any** parametrization. In proving this result it was necessary to be careful in our use of the variable t. At certain times it denoted a point in the interval $[0, l]$, the domain of the unit speed parametrization $P \circ s^{-1}$, but in our final formula for $k(t)$ it denotes a point in the interval $[a, b]$—the domain of definition of P.

In the real world we are usually presented with a curve in its geometric form, i.e., as a set of points, and in order to use the theory we have developed it is necessary to parametrize these geometric objects. If the curve is not closed it is first necessary to choose an initial and final point in order to obtain a directed curve. We look at these problems, beginning with the known standard parametrization of the unit circle $x^2 + y^2 = 1$, $P(t) = (\cos t, \sin t)$, $0 \le t \le 2\pi$. Since this is a parametrization we have

$$\cos^2 t + \sin^2 t = 1.$$

So we have two very similar formulae

$$x^2 + y^2 = 1$$

and

$$\cos^2 t + \sin^2 t = 1.$$

Recognizing the similarity between these formulae would have led us to the parametrization and, in general, if we recognize an equation involving a function of a **single** variable which is similar to the equation satisfied by the level set then we can write down a parametrization. Do not forget that it is also necessary to write down the domain of the parametrization and to check that the sense of direction is correct.

Example 54. Consider the **ellipse**

$$\frac{x^2}{a^2} + \frac{y^2}{b^2} = 1$$

where $a > b > 0$. Our equation is

$$\left(\frac{x}{a}\right)^2 + \left(\frac{y}{b}\right)^2 = 1$$

and we compare this with

$$\cos^2 t + \sin^2 t = 1.$$

If we let $\dfrac{x}{a} = \cos t$ and $\dfrac{y}{b} = \sin t$ we get an anticlockwise parametrization
$$P(t) = \big(x(t), y(t)\big) = (a\cos t, b\sin t)$$
where $0 \leq t \leq 2\pi$. For the curvature of the ellipse we obtain
$$k(t) = \frac{(-a\sin t)(-b\sin t) - (b\cos t)(-a\cos t)}{\big[(-a\sin t)^2 + (b\cos t)^2\big]^{3/2}}$$
$$= \frac{ab}{(a^2 \sin^2 t + b^2 \cos^2 t)^{3/2}}.$$

If we take an ellipse with $a = b$ we get a circle of radius a with centre at the origin and this observation allows us an opportunity to check a special case of our formula. In this case the curvature reduces to
$$\frac{ab}{(a^2 \sin^2 t + b^2 \cos^2 t)^{3/2}} = \frac{a \cdot a}{(a^2 \sin^2 t + a^2 \cos^2 t)^{3/2}} = \frac{a^2}{(a^2)^{3/2}} = \frac{1}{a}$$
which agrees with our earlier observation.

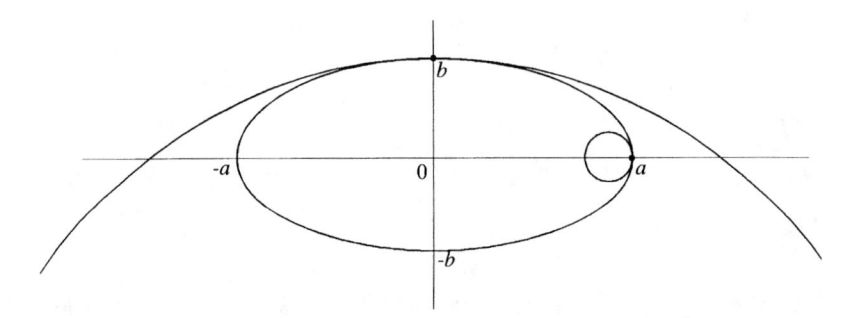

Figure 14.2

We shall soon see that the absolute curvature at a point may be interpreted as the reciprocal of the radius of the circle which sits closest to the curve at the point. By inspection of Figure 14.2 we see that the absolute curvature is a minimum at the points $(0, b)$ and $(0, -b)$ and a maximum at the points $(a, 0)$ and $(-a, 0)$. This can be verified by finding the maximum and minimum of $k(t)$ and is also obvious by inspection of the formula for $k(t)$.

The maximum curvature is $\dfrac{ab}{(b^2)^{3/2}} = \dfrac{a}{b^2}$ and the minimum curvature is $\dfrac{ab}{(a^2)^{3/2}} = \dfrac{b}{a^2}$. We have minimum curvature = maximum curvature if and only if $\dfrac{a}{b^2} = \dfrac{b}{a^2}$, i.e., if and only if $a^3 = b^3$ and hence $a = b$ and we have proved that an ellipse has constant curvature if and only if it is a circle.

Curvature

The ancient Greeks also developed Euclidean geometry in three-dimensional space and in this manner discovered the ellipse. They considered a cone, which is constructed using a straight line and circle, and then took cross sections (Figure 14.3), as we did earlier in moving from two to three dimensions. These cross sections are known as **conic sections** and yield the ellipse, the hyperbola and the parabola.

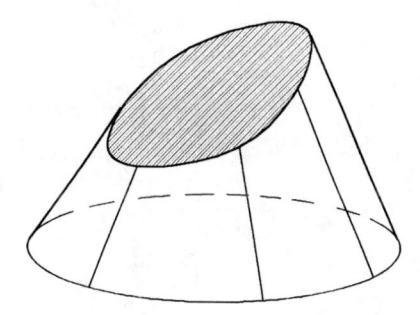

Figure 14.3

Example 55. In this example we consider the hyperbola

$$\frac{x^2}{a^2} - \frac{y^2}{b^2} = 1.$$

If $x = 0$ then $\frac{-y^2}{b^2} = 1$ and this is impossible so the level set never cuts the y-axis. Figure 14.4 shows that the hyperbola consists of two parts. By the intermediate value theorem the image of an interval by a continuous function cannot consist of two parts. This rules out the possibility of one parametrization for the complete hyperbola. We proceed by considering each part in turn. We have another problem, as neither part is a closed curve, and it is not clear how to choose an initial and final point. We can, however, give each part a sense of direction as indicated in Figure 14.4. Let us proceed as we did in the previous example and see what happens.

We recall the definition of the **hyperbolic functions**

$$\cosh x = \frac{e^x + e^{-x}}{2} \quad \text{and} \quad \sinh x = \frac{e^x - e^{-x}}{2}.$$

These have very similar properties to the **trigonometric** functions since

$$\cos x = \frac{e^{ix} + e^{-ix}}{2} \quad \text{and} \quad \sin x = \frac{e^{ix} - e^{-ix}}{2i}.$$

In particular we have the easily verifiable properties

$$\frac{d}{dx}(\sinh x) = \cosh x \quad \text{and} \quad \frac{d}{dx}(\cosh x) = \sinh x$$

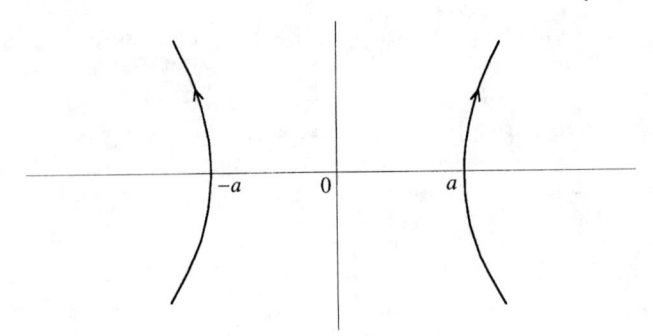

Figure 14.4

and
$$\cosh^2 x - \sinh^2 x = 1.$$

Comparing this last formula with
$$\left(\frac{x}{a}\right)^2 - \left(\frac{y}{b}\right)^2 = 1$$

gives
$$P(t) = (a \cosh t, b \sinh t).$$

The mapping P, restricted to $[a, b]$, parametrizes a part of the curve on the right-hand side of Figure 14.4 and if we allow a parametrization using intervals of infinite length then P with domain $(-\infty, +\infty)$ parametrizes the full curve on the right-hand side.

Observing that
$$\left(\frac{-x}{a}\right)^2 - \left(\frac{y}{b}\right)^2 = 1$$

we see that $Q(t) = (-a \cosh t, b \sinh t)$ parametrizes the curve on the left-hand side. From the curvature formula we obtain, for the right-hand curve,
$$k(t) = \frac{b \sinh(t) a \sinh(t) - a \cosh(t) b \cosh(t)}{(a^2 \sinh^2 t + b^2 \cosh^2 t)^{3/2}}$$
$$= \frac{-ab}{(a^2 \sinh^2 t + b^2 \cosh^2 t)^{3/2}}.$$

This example showed up limitations in our definition of directed curve. Our definition could be modified to include this level set, or at least each part of it separately. We prefer, however, to stay with our original definition and to handle situations like this on a case by case basis rather than developing a more complicated theory. This is a typical example of how mathematics is developed and research undertaken. A simple theory is studied and developed and when sufficiently many natural examples do not fit into the

Curvature

theory one searches for a more general theory in order to include these examples.

In many situations, there is no obvious way to find a parametrization, but surprisingly, it is rather easy to parametrize the graph of a function of one variable. This is our next example.

Example 56. We suppose that the curve Γ is the graph of a twice continuously differentiable function f from $\mathbf{R}$ into $\mathbf{R}$. The points on the graph are precisely the points of the form $(x, f(x))$ and so the mapping

$$[a,b] \to \Gamma$$
$$t \to (t, f(t))$$

is a parametrized curve. Let $P(t) = (x(t), y(t)) = (t, f(t))$. We have $P'(t) = (1, f'(t)) \neq (0,0)$ for all t. Substituting into the formula for curvature gives

$$k(t) = \frac{1 \cdot f''(t) - 0 \cdot f'(t)}{\left(1 + (f'(t))^2\right)^{3/2}}, \quad \text{i.e.,} \quad k = \frac{f''}{\left(1 + (f')^2\right)^{3/2}}.$$

Note that we did not begin this example by specifying an initial and final point on the graph. By accepting a parametrization we also implicitly accepted an initial and final point. Unless we state otherwise we always use this sense of direction, which is **inherited** from the **positive** direction of the x-axis, on a graph.

In our discussion on Lagrange multipliers we noted that locally every level set is the graph of a function. If we can find this function then we may apply the result of Example 56. We take this approach in our next example.

Example 57. We parametrize once more the unit circle $x^2 + y^2 = 1$. If $x = t$ then $y^2 = 1 - t^2$ and $y = \pm\sqrt{1 - t^2}$. We have two solutions each of which gives a parametrization of a part of the unit circle. If we take the positive square root, then $y > 0$ and

$$P_1(t) = \left(-t, \sqrt{1 - t^2}\right), \quad -1 \le t \le 1,$$

parametrizes the upper semicircle (Figure 14.5) while $P_2(t) = \left(t, -\sqrt{1 - t^2}\right)$, $-1 \le t \le 1$, parametrizes the lower semicircle and we obtain enough parametrizations to calculate the curvature at almost all points.

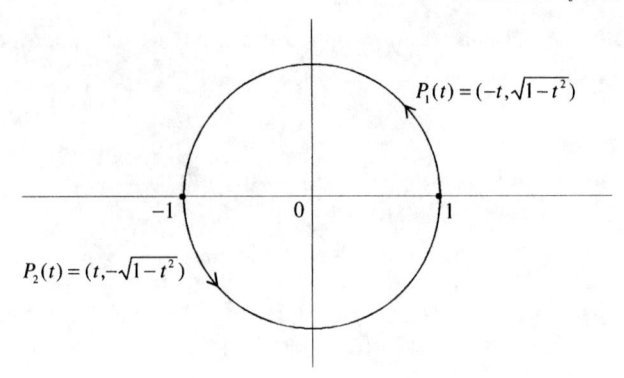

Figure 14.5

$P_1(t) = (-t, \sqrt{1-t^2})$
$P_2(t) = (t, -\sqrt{1-t^2})$

This approach is suitable if we can solve a certain equation but this may often be very difficult. The same problem arose in our discussion on Lagrange multipliers and we use the same methods here to find a solution. We will now find the curvature of a level set without recourse to a parametrization. Let $\Gamma = \{(x, y); f(x, y) = 0\}$ and suppose $\nabla f(x, y) \neq (0, 0)$ at all points of Γ. We will use the notation f_x in place of $\dfrac{\partial f}{\partial x}$, f_y in place of $\dfrac{\partial f}{\partial y}$, f_{xy} in place of $\dfrac{\partial^2 f}{\partial x \partial y}$ etc.

Let $(a, b) \in \Gamma$ and suppose $f_y(a, b) \neq 0$. By the implicit function theorem there exists a function ϕ defined near a such that $\phi(a) = b$ and the level set near (a, b) coincides with the graph of ϕ. Hence

$$f(x, \phi(x)) = 0 \qquad (*)$$

and, by Example 56, the curvature of the level set near (a, b), with direction inherited from the graph, is

$$\frac{\phi''}{\left(1 + (\phi')^2\right)^{3/2}}.$$

Using $(*)$ as in Chapter 8, or Example 50, we obtain

$$\phi'(x) = -\frac{f_x}{f_y}.$$

Curvature

A further application of the chain rule yields

$$\phi''(x) = \frac{-[(f_{xx} + f_{xy}\phi')f_y - (f_{xy} + f_{yy}\phi')f_x]}{f_y^2}$$

$$= \frac{-\left[(f_{xx} - f_{xy}\frac{f_x}{f_y})f_y - (f_{xy} - f_{yy}\frac{f_x}{f_y})f_x\right]}{f_y^2}$$

$$= \frac{-[f_{xx}f_y^2 - 2f_{xy}f_xf_y + f_{yy}f_x^2]}{f_y^3}.$$

Hence

$$k = \frac{\phi''}{(1+(\phi')^2)^{3/2}} = \frac{-f_{xx}f_y^2 + 2f_{xy}f_xf_y - f_{yy}f_x^2}{f_y^3\left[1+\left(\frac{f_x}{f_y}\right)^2\right]^{3/2}}.$$

In simplifying this expression do not forget that $(f_y^2)^{3/2} = |f_y|^3$. This explains the $(\pm)$ sign in the final expression

$$k = \pm\frac{-f_{xx}f_y^2 + 2f_{xy}f_xf_y - f_{yy}f_x^2}{|f_x^2 + f_y^2|^{3/2}}$$

where we take the positive sign if $f_y > 0$ and the negative sign if $f_y < 0$. If $f_x \neq 0$ we arrive at a similar formula.

Example 58. If $f(x,y) = \dfrac{x^2}{a^2} + \dfrac{y^2}{b^2} - 1$, then the ellipse is obtained as the level set

$$\Gamma = \{(x,y); f(x,y) = 0\}.$$

On computing derivatives we get

$$f_x = \frac{2x}{a^2}, \qquad f_y = \frac{2y}{b^2}$$

$$f_{xx} = \frac{2}{a^2}, \qquad f_{yy} = \frac{2}{b^2}$$

$$f_{xy} = f_{yx} = 0.$$

Confining ourselves to the upper portion of the ellipse, we have $f_y > 0$

and

$$k = \text{curvature at the point } (x, y)$$

$$= \frac{-\frac{2}{a^2}\left(\frac{2y}{b^2}\right)^2 - \frac{2}{b^2}\left(\frac{2x}{a^2}\right)^2}{\left|\left(\frac{2x}{a^2}\right)^2 + \left(\frac{2y}{b^2}\right)^2\right|^{3/2}}$$

$$= \frac{-\frac{1}{a^2b^2}\left(\frac{y^2}{b^2} + \frac{x^2}{a^2}\right)}{\frac{1}{a^3b^3}\left|\frac{a^2b^2x^2}{a^4} + \frac{a^2b^2y^2}{b^4}\right|^{3/2}}$$

$$= \frac{-ab}{\left|\frac{b^2x^2}{a^2} + \frac{a^2y^2}{b^2}\right|^{3/2}}.$$

We compare this with our earlier form of the curvature (Example 54) by making the substitution $x = a\cos t$, $y = b\sin t$. We obtain

$$k(t) = \frac{-ab}{\left|\frac{b^2a^2\cos^2 t}{a^2} + \frac{a^2b^2\sin^2 t}{b^2}\right|^{3/2}}$$

$$= \frac{-ab}{\left(b^2\cos^2 t + a^2\sin^2 t\right)^{3/2}}$$

which is, apart from the change of sign, the formula we obtained previously. The change of sign arises from the fact that the parametrization coming from the graph is $t \to \left(t, b\sqrt{1 - \frac{t^2}{a^2}}\right)$, $-1 \le t \le +1$, and hence the initial and final points are $(-a, 0)$ and $(a, 0)$, respectively. This is allowed since the graph is not a closed curve. However, the full ellipse is closed and hence directed in the opposite, i.e., anticlockwise, direction.

In this example we were able to calculate the curvature directly from the partial derivatives of the function defining the level set—we did not have to obtain a parametrization and we did not have to display the level set as the graph of an explicit function. We also learned that it is necessary to check that the parametrization has the right sense of direction. In Chapter 16 we shall see how the sign of the curvature and some idea of the shape of Γ help us to distinguish between parametrizations of a directed curve and parametrizations of the same curve directed in the opposite way.

Curvature

Exercises

14.1 Find the absolute curvature of the graph of $f(x) = e^x$ and the level set $x^3 - y^3 = 2$.

14.2 We have $\cos(x) = \dfrac{e^{ix} + e^{-ix}}{2}$, $\cosh(x) = \dfrac{e^x + e^{-x}}{2}$. Show $\cosh(ix) = \cos x$ and $\cos(ix) = \cosh x$. By differentiating find similar formulae for $\sin x$ and $\tan x$. Using $\cosh$ and $\sinh$ parametrize the curve $\dfrac{x^2}{16} - \dfrac{y^2}{81} = 1$ and find its curvature.

14.3 Parameterize the curve $\dfrac{x^2}{9} + y^2 = 9$ and hence find its curvature. Find the points where the curvature is a maximum (a) by inspection of the level curve, (b) by differentiating the curvature function, (c) by inspection of the curvature function.

14.4 Show that the curvature of $P(t) = (t, \log \cos t)$, $-\frac{\pi}{2} < t < \frac{\pi}{2}$, is $-\cos t$.

14.5 Do Exercises 14.2 and 14.3 again using partial derivatives.

14.6 Find the closest points on the curve $x^2 - y^2 = 1$ to $(a, 0)$ where (i) $a = 4$ (ii) $a = 2$ (iii) $a = \sqrt{2}$.

14.7 Find the curvature of the closed curve defined by the equation
$$9x^2 + 4y^2 - 36x - 24y + 71 = 0.$$
By completing squares show that this curve describes an ellipse. Parametrize the ellipse and using the parametrization find its curvature.

14.8 Let c denote a positive real number. Show that the set of all $(x, y) \in \mathbf{R}^2$ satisfying
$$d((x, y), (c, 0)) + d((x, y), (-c, 0)) = 3c$$
is an ellipse (d denotes distance).

14.9 Parameterize the graph of $f(x) = x(1 - x)$, $0 \le x \le 1$, and find the points on the graph where the curvature is a maximum and a minimum.

14.10 Show that a directed curve in $\mathbf{R}^2$ is a straight line if and only if all its tangent lines are parallel.

15

Quadratic Approximation

Summary. *We obtain a quadratic approximation for a twice differentiable function of one variable. Using an estimate of the error we complete the proof of Proposition 21 and also show that absolute curvature at a point on a curve may be interpreted as the reciprocal of the radius of the circle which lies closest to the curve at that point.*

In Chapter 7 we used linear approximations to show

$$\frac{\partial f}{\partial \vec{v}} = v_1 \frac{\partial f}{\partial x} + v_2 \frac{\partial f}{\partial y}.$$

In this chapter we approximate functions of one variable by using the first and second derivatives and apply this to complete the proof of Proposition 21 and to give a geometrical interpretation of absolute curvature. The details are rather technical but cannot be avoided. In our case we decided to allow the methods of linear approximation to be absorbed before moving to quadratic approximation. We begin by recalling the one variable linear case. If ϕ is differentiable, then

$$\phi(x + \Delta x) = \phi(x) + \phi'(x) \cdot \Delta x + g(x, \Delta x) \cdot \Delta x$$

where $g(x, \Delta x) \to 0$ as $\Delta x \to 0$.

At the time we asked what happens when we take more derivatives. Well, we need to know the answer now. We assume ϕ'' exists and apply the above formula to $\phi'(x)$ to get

$$\phi'(x + \Delta x) = \phi'(x) + \phi''(x) \cdot \Delta x + h(x, \Delta x) \cdot \Delta x$$

where $h(x, \Delta x) \to 0$ as $\Delta x \to 0$. Since ϕ' is continuous it follows that, for

Quadratic Approximation

fixed x, $h(x, \Delta x)$ is a continuous function of Δx and hence is integrable. If we replace Δx by s and integrate with respect to s we get

$$\int_0^{1} {}^x \phi'(x+s)\, ds = \int_0^{1} {}^x \phi'(x)\, ds + \int_0^{1} {}^x s\phi''(x)\, ds + \int_0^{1} {}^x sh(x,s)\, ds.$$

By the fundamental theorem of calculus

$$\int_0^{1} {}^x \phi'(x+s)\, ds = \phi(x + \Delta x) - \phi(x)$$

so

$$\phi(x + \Delta x) - \phi(x) = \phi'(x) \cdot \Delta x + \phi''(x) \frac{(\Delta x)^2}{2} + \int_0^{1} {}^x sh(x,s)\, ds.$$

Let $l(x, \Delta x) \cdot (\Delta x)^2 = \int_0^{1} {}^x sh(x,s)\, ds$.

We now apply the following result from the one variable integral calculus: if f is a continuous function on $[a, b]$, then

$$\left| \int_a^b f(x)\, dx \right| \leq (b - a) \max\{|f(x)|; a \leq x \leq b\}.$$

Hence

$$\left| \int_0^{1} {}^x sh(x,s)\, ds \right| \leq |\Delta x| \max\{|sh(x,s)|; 0 \leq s \leq \Delta x\}$$

$$\leq (\Delta x)^2 \max\{|h(x,s)|; 0 \leq s \leq \Delta x\}.$$

Since $h(x, \Delta x) \to 0$ as $\Delta x \to 0$ it follows that we have shown

$$\phi(x + \Delta x) = \phi(x) + \phi'(x) \cdot \Delta x + \frac{\phi''(x)}{2}(\Delta x)^2 + l(x, \Delta x) \cdot (\Delta x)^2$$

where $l(x, \Delta x) \to 0$ as $\Delta x \to 0$ and obtained the **quadratic approximation** to $\phi(x + \Delta x)$,

$$\phi(x) + \phi'(x) \cdot \Delta x + \frac{\phi''(x)}{2}(\Delta x)^2.$$

If, in addition, ϕ'' exists and is continuous near x then, by the mean value theorem,

$$h(x, s) = \frac{\phi'(x+s) - \phi'(x)}{s} - \phi''(x)$$

$$= \phi''(x + \theta s) - \phi''(x)$$

for some θ, $0 < \theta < 1$, and we have the improved estimate

$$|l(x, \Delta x)| \leq \sup\{|h(x,s)|; 0 \leq s \leq \Delta x\}$$

$$\leq \sup_{|\theta| \leq 1} |\phi''(x + \theta \cdot \Delta x) - \phi''(x)|.$$

We will apply the above to interpret curvature but first we apply both the estimate and the error to complete the proof of Proposition 21. We recall the setting: the function f, defined on an open subset of $\mathbf{R}^2$, has continuous first and second order partial derivatives at a critical point which we suppose for the sake of simplicity and without loss of generality, is the origin in $\mathbf{R}^2$, $(0,0)$. Let $A = \frac{\partial^2 f}{\partial x^2}(0,0)$, $B = \frac{\partial^2 f}{\partial x \partial y}(0,0)$ and $C = \frac{\partial^2 f}{\partial y^2}(0,0)$. We consider only the case of a local minimum; the local maximum case is handled in the same way. We suppose

$$Av_1^2 + 2Bv_1 v_2 + Cv_2^2 > 0$$

for all non-zero (v_1, v_2) in $\mathbf{R}^2$. We have seen, in Chapter 4, that this implies there exists $\alpha > 0$ such that

$$Av_1^2 + 2Bv_1 v_2 + Cv_2^2 \geq \alpha(v_1^2 + v_2^2) \qquad (1)$$

for all $(v_1, v_2) \in \mathbf{R}^2$. Let $h(t) = f(t\vec{v})$. By the chain rule

$$h'(t) = \frac{d}{dt}(f(t\vec{v})) = \nabla f(t\vec{v}) \cdot \vec{v} = \frac{\partial f}{\partial \vec{v}}(t\vec{v}).$$

Applying the same rule to $\frac{\partial f}{\partial \vec{v}}(t\vec{v})$ we get

$$h''(t) = \frac{d^2}{dt^2}(f(t\vec{v})) = \frac{\partial^2 f}{\partial \vec{v}^2}(t\vec{v}).$$

We apply the quadratic approximation to h and get, since $(0,0)$ is a critical point of f,

$$f(t\vec{v}) = f(0,0) + \frac{\partial^2 f}{\partial \vec{v}^2}(0,0) \cdot \frac{t^2}{2} + l(0,0,\vec{v},t) \cdot t^2 \qquad (2)$$

where

$$|l(0,0,\vec{v},t)| \leq \sup_{|\theta| \leq |t|} \left| \frac{\partial^2 f}{\partial \vec{v}^2}(\theta \vec{v}) - \frac{\partial^2 f}{\partial \vec{v}^2}(0,0) \right|.$$

Now

$$\frac{\partial^2 f}{\partial \vec{v}^2}(\theta \vec{v}) - \frac{\partial^2 f}{\partial \vec{v}^2}(0,0)$$
$$= (f_{xx}(\theta \vec{v}) - A)v_1^2 + 2(f_{xy}(\theta \vec{v}) - B)v_1 v_2 + (f_{yy}(\theta \vec{v}) - C)v_2^2$$

and, as the second order partial derivatives of f are continuous, we can find $\delta > 0$ such that

$$|f_{xx}(\theta \vec{v}) - A| \leq \frac{\alpha}{8}, |f_{xy}(\theta \vec{v}) - B| \leq \frac{\alpha}{8} \text{ and } |f_{yy}(\theta \vec{v}) - C| \leq \frac{\alpha}{8}$$

for all $|\theta| \leq \delta$ and all $\vec{v}$, $\|\vec{v}\| = 1$.

Quadratic Approximation

Hence, using $|2v_1 v_2| \leq v_1^2 + v_2^2$, we have for $\|\vec{v}\| = 1$ and $|t| \leq \delta$

$$|l(0,0,\vec{v},t)| \leq \frac{\alpha}{8}v_1^2 + 2 \cdot \frac{\alpha}{8}|v_1 v_2| + \frac{\alpha}{8}v_2^2$$
$$\leq \frac{\alpha}{8}v_1^2 + \frac{\alpha}{8}(v_1 + v_2) + \frac{\alpha}{8}v_2^2 \qquad (3)$$
$$= \frac{\alpha}{4}(v_1 + v_2)$$
$$= \frac{\alpha}{4}.$$

Combining (1), (2) and (3) we obtain

$$f(t\vec{v}) \geq f(0,0) + \alpha \frac{t^2}{2} - \frac{\alpha}{4}t^2 > f(0,0)$$

for all t, $0 < |t| \leq \delta$, and all $\vec{v}$, $\|\vec{v}\| = 1$. Since

$$\{t\vec{v}; \quad |t| \leq \delta, \quad \|\vec{v}\| = 1\}$$

is a disc of positive radius centered at the origin, we have shown that $(0,0)$ is a local minimum. This completes the proof of Proposition 2.1.

We now return to curvature and suppose we are given a curve Γ. To simplify matters we suppose that Γ is the graph of a function and that the curvature is non-zero. The radius of the circle which sits closest to the curve near a given point A is unchanged if we change the origin in $\mathbf{R}^2$ or rotate the axes. We choose an origin and axis to suit ourselves. We place the origin at A and the x-axis in the direction of the tangent at A (Figure 15.1). Turning the page around we get Figure 15.2.

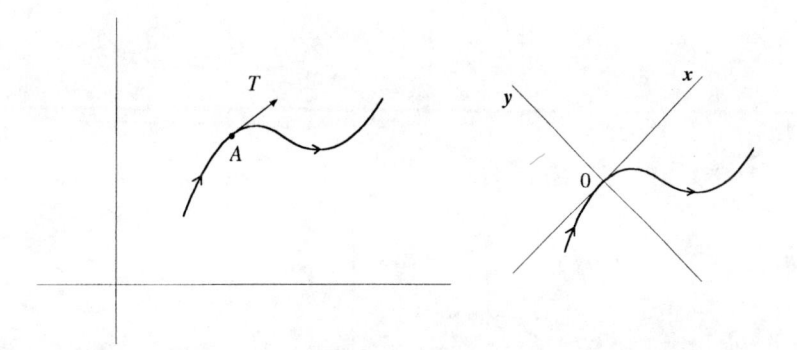

Figure 15.1

If our curve is the graph of the function ϕ, then $\phi(0) = 0$ and $\phi'(0) = 0$. Since $\phi''(0)$ exists we have

$$\phi(\Delta x) = \frac{\phi''(0)(\Delta x)^2}{2} + k(x, \Delta x) \cdot (\Delta x)^2$$

Functions of Two Variables

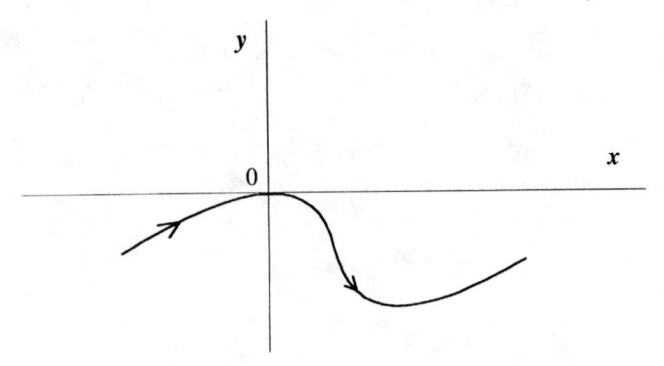

Figure 15.2

where $k(x, \Delta x) \to 0$ as $\Delta x \to 0$. The curvature at the origin is

$$\frac{\phi''(0)}{(1+\phi'(0)^2)^{3/2}} = \phi''(0).$$

We wish to find the circle closest to the curve at the origin. Through any three distinct points which do not lie on a straight line one can draw precisely one circle (see Exercise 15.4). Let (Figure 15.3)

$$P = \left(-\Delta x, \frac{\phi''(0) \cdot (\Delta x)^2}{2} + k(x, -\Delta x) \cdot (\Delta x)^2\right)$$

and

$$Q = \left(\Delta x, \frac{\phi''(0) \cdot (\Delta x)^2}{2} + k(x, \Delta x) \cdot (\Delta x)^2\right).$$

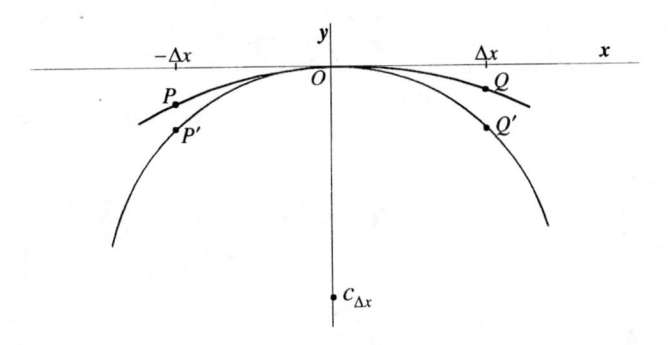

Figure 15.3

The limit of the circles going through the points POQ as $\Delta x \to 0$ will be the circle we are seeking—the **circle of curvature**. Since $k(x, \Delta x)$ and $k(x, -\Delta x)$ tend to zero as $\Delta x \to 0$ and $\phi''(0) \neq 0$ it suffices to take the

Quadratic Approximation

limit of circles through the points $P' = \left(-\Delta x, \dfrac{\phi''(0) \cdot (\Delta x)^2}{2}\right)$, $O = (0,0)$ and $Q' = \left(\Delta x, \dfrac{\phi''(0) \cdot (\Delta x)^2}{2}\right)$ as $\Delta x \to 0$. Let $c_{1\,x}$ and $R_{1\,x}$ denote the centre and radius of this circle. Since the y-axis is the line equidistant from P' and Q' it follows that $c_{1\,x}$ is on the y-axis. Since the origin is also on the circle we have

$$c_{1\,x} = (0, -R_{1\,x}).$$

Hence

$$\text{distance}^2(c_{1\,x}, O) = \text{distance}^2(c_{1\,x}, P')$$

and

$$R_{1\,x}^2 = (\Delta x)^2 + \left(-R_{1\,x} - \dfrac{\phi''(0) \cdot (\Delta x)^2}{2}\right)^2$$

$$= (\Delta x)^2 + R_{1\,x}^2 + 2R_{1\,x} \dfrac{\phi''(0)}{2}(\Delta x)^2 + \dfrac{(\phi''(0))^2}{4}(\Delta x)^4.$$

On simplifying we obtain

$$(\Delta x)^2 = -R_{1\,x} \cdot \phi''(0)(\Delta x)^2 - \dfrac{(\phi''(0))^2}{4}(\Delta x)^4.$$

Dividing both sides by $(\Delta x)^2$ gives

$$1 = -R_{1\,x} \cdot \phi''(0) - \dfrac{(\phi''(0))^2}{4}(\Delta x)^2$$

and letting $\Delta x \to 0$ we get

$$-1 = \left(\lim_{1\ x \to 0} R_{1\,x}\right) \phi''(0).$$

If $r = \lim\limits_{1\ x \to 0} R_{1\,x}$ then we have shown that

$$|\phi''(0)| = \dfrac{1}{r}.$$

This establishes that the absolute curvature is the reciprocal of the radius of the circle that sits closest to the curve. The centre of this circle is called the **centre of curvature** and we have seen that it lies on the y-axis. This shows that in general the centre lies on the line perpendicular to the tangent, i.e., on the **normal**. Moreover, the minus sign appeared because the normal and the circle of curvature were on opposite sides of the curve.

In the above analysis we have made quite a number of assumptions. Can you identify them? Can you justify them? For instance we used P' and Q' in place of P and Q mainly for the sake of having simpler calculations. If you rewrite the above with P and Q in place of P' and Q' you will find that you get the same answer—this justifies our use of P' and Q'. We choose a special point as the origin and a certain direction as the x-axis. It would be rather remarkable if the curvature depended on this choice. Nevertheless, you have only my word that it doesn't. How do you justify this assumption? The most direct way is to take an arbitrary point and to suppose the tangent points in any non-vertical direction and to rewrite the proof once more with this modification. Alternatively, one can show that curvature is unchanged under a certain "change of variables". We discuss this idea in Chapter 21 but do not consider this particular problem. Again you would arrive at the same conclusion. Why do we not allow a vertical tangent? Since there are curves with vertical tangents, for instance a complete circle, how do we remove this restriction? We assumed that the curvature was non-zero and that our curve was part of a level set. Again the same proof can be developed to take care of any parametrized curve with non-zero velocity at all points. Alternatively one can show that all such curves are in fact locally level sets.

This process of starting with a proof in a very nice situation and afterwards refining the proof to deal with less regular situations is extremely common in mathematics. Unfortunately, it is also extremely common to provide merely the final complicated proof in print. This often hides the intuitive ideas behind the proof—which are often as important as the careful details of the final version. Try working out this programme and you will really learn something.

Exercises

15.1 Work out a quadratic approximation for functions of two variables and use it to improve Exercise 7.2.

15.2 Particles A and B travel along elliptical paths with position at time t given by $P_A(t) = (4\cos t, 2\sin t)$ and $P_B(t) = (2\sin 2t, 3\cos 2t)$, $t \geq 0$. Sketch the paths followed by each particle. Use the chain rule to find the rate at which the distance between the objects is changing at $t = \pi/2$.

15.3 Let P, Q and R denote three distinct points in $\mathbf{R}^2$ which do not lie on a straight line. Show that the perpendicular bisectors of the line segments $[PQ]$, $[QR]$ and $[RP]$ meet at a point c which is equidistant from P, Q and R. Hence show that c is the centre of a circle which passes through P, Q and R.

16

Vector Valued Differentiation

Summary. *We prove product rules for differentiating vector valued functions. We derive a relationship between curvature, the derivative of the tangent and the normal. We interpret the sign of curvature and state a result about local maxima and minima of the curvature.*

One of the most useful techniques in the one-dimensional differential calculus is the **product rule**

$$\frac{d}{dx}(fg) = \frac{df}{dx}g + f\frac{dg}{dx}.$$

We look at variations of this rule for functions from $\mathbf{R}$ into $\mathbf{R}^2$ and $\mathbf{R}^2$ into $\mathbf{R}$.

First we consider functions $f, g : U \subset \mathbf{R}^2 \to \mathbf{R}$ which are differentiable. Since we have already found that ∇ behaves like $\dfrac{d}{dx}$ let us calculate $\nabla(fg)$. We have

$$\nabla(fg) = \left(\frac{\partial}{\partial x}(fg), \frac{\partial}{\partial y}(fg)\right)$$

$$= \left(\frac{\partial f}{\partial x}g + f\frac{\partial g}{\partial x}, \frac{\partial f}{\partial y}g + f\frac{\partial g}{\partial y}\right)$$

$$= \left(\frac{\partial f}{\partial x}g, \frac{\partial f}{\partial y}g\right) + \left(f\frac{\partial g}{\partial x}, f\frac{\partial g}{\partial y}\right)$$

$$= g\left(\frac{\partial f}{\partial x}, \frac{\partial f}{\partial y}\right) + f\left(\frac{\partial g}{\partial x}, \frac{\partial g}{\partial y}\right)$$

$$= g\nabla f + f\nabla g.$$

So $\nabla(fg) = g\nabla f + f\nabla g$ and this is a product rule for differentiating $\mathbf{R}$-valued functions of two variables.

Example 59. Let $f(x,y) = e^{xy}$ and $g(x,y) = x^2 + y^2$. Then
$$\nabla f = (ye^{xy}, xe^{xy})$$
and
$$\nabla g = (2x, 2y).$$
Since $fg(x,y) = (x^2 + y^2)e^{xy}$ we have
$$\nabla(fg) = (2xe^{xy} + (x^2+y^2)ye^{xy}, 2ye^{xy} + (x^2+y^2)xe^{xy}).$$
On the other hand
$$\begin{aligned}g\nabla f + f\nabla g &= (x^2+y^2)(ye^{xy}, xe^{xy}) + e^{xy}(2x, 2y)\\ &= ((x^2+y^2)ye^{xy}, (x^2+y^2)xe^{xy}) + (e^{xy}2x, e^{xy}2y)\\ &= ((x^2+y^2)ye^{xy} + 2xe^{xy}, (x^2+y^2)xe^{xy} + 2ye^{xy})\\ &= \nabla(fg).\end{aligned}$$

Now consider functions from $\mathbf{R}$ into $\mathbf{R}^2$. Let $\phi(t) = \big(x_1(t), y_1(t)\big)$ and $\psi(t) = \big(x_2(t), y_2(t)\big)$. If we take the dot product of ϕ and ψ we get a real valued function
$$\begin{aligned}(\phi \cdot \psi)(t) &= \big(x_1(t), y_1(t)\big) \cdot \big(x_2(t), y_2(t)\big)\\ &= x_1(t)x_2(t) + y_1(t)y_2(t).\end{aligned}$$
We have
$$\begin{aligned}\frac{d}{dt}(\phi \cdot \psi)(t) &= \frac{d}{dt}\big(x_1(t)x_2(t)\big) + \frac{d}{dt}\big(y_1(t)y_2(t)\big)\\ &= x_1'(t)x_2(t) + x_1(t)x_2'(t) + y_1'(t)y_2(t) + y_1(t)y_2'(t)\\ &= \big(x_1'(t), y_1'(t)\big) \cdot \big(x_2(t), y_2(t)\big) + \big(x_1(t), y_1(t)\big) \cdot \big(x_2'(t), y_2'(t)\big)\\ &= \phi'(t) \cdot \psi(t) + \phi(t) \cdot \psi'(t)\,,\end{aligned}$$
i.e.,
$$(\phi \cdot \psi)' = \phi' \cdot \psi + \phi \cdot \psi'.$$
Again we have a rule for differentiation but now for the dot product.

Vector valued differentiation is an important tool in the study of curves in $\mathbf{R}^3$. We briefly discuss the situation for plane curves, as even here it gives us some insight.

Let $P : [a,b] \to \mathbf{R}^2$ denote a unit speed parametrized curve. Then
$$\|P'(t)\| = 1 \quad \text{for all } t.$$
Hence $\|P'(t)\|^2 = 1$ for all t and so $P'(t) \cdot P'(t) = 1$ where $\cdot$ is the dot

Vector Valued Differentiation 127

product. Hence $P'(t)$ is the unit tangent $T(t)$ and $\frac{d}{dt}(T(t) \cdot T(t)) = 0$. By the dot product rule for differentiating we have

$$T'(t) \cdot T(t) + T(t) \cdot T'(t) = 0.$$

Since the dot product of vectors $\vec{a}$ and $\vec{b}$ in $\mathbf{R}^2$ has the property that $\vec{a} \cdot \vec{b} = \vec{b} \cdot \vec{a}$ it follows that

$$2T'(t) \cdot T(t) = 0 \quad \text{and} \quad T'(t) \cdot T(t) = 0.$$

As two vectors are perpendicular if and only if their dot product is zero we have $T'(t) \perp T(t)$ and since we are considering vectors in $\mathbf{R}^2$ it follows that $T'(t)$ must point in the direction of the unit normal $N(t)$. Hence there is some real number that depends only on t—and so we write $\alpha(t)$—such that

$$T'(t) = \alpha(t)N(t).$$

Taking the dot product with $N(t)$ we get

$$T'(t) \cdot N(t) = \alpha(t)N(t) \cdot N(t) = \alpha(t).$$

If $P(t) = (x(t), y(t))$ then

$$P'(t) = T(t) = (x'(t), y'(t)),$$
$$T'(t) = (x''(t), y''(t))$$
$$N(t) = (-y'(t), x'(t)).$$

Hence

$$T'(t) \cdot N(t) = (x''(t), y''(t)) \cdot (-y'(t), x'(t))$$
$$= -x''(t)y'(t) + y''(t)x'(t)$$
$$= \kappa(t)$$

by the formula for the curvature of unit speed parametrized curves established in Chapter 14. Hence

$$T'(t) = \kappa(t)N(t). \qquad (1)$$

Since $N(t) \cdot N(t) = 1$ it follows also that

$$N'(t) \cdot N(t) = 0$$

and $N'(t)$ is parallel to $T(t)$. This shows that

$$N'(t) = \beta(t)T(t). \qquad (2)$$

Moreover, since $T(t) \cdot N(t) = 0$ we have, on differentiating,

$$T'(t) \cdot N(t) + T(t) \cdot N'(t) = 0. \qquad (3)$$

Substituting (1) and (2) into (3) gives

$$\kappa(t)N(t) \cdot N(t) + \beta(t)T(t) \cdot T(t) = 0,$$

i.e., $\kappa(t) + \beta(t) = 0$ and we have proved

$$N'(t) = -\kappa(t)T(t). \tag{4}$$

Equations (1) and (4) are often written in matrix form,

$$\begin{pmatrix} T \\ N \end{pmatrix}' = \begin{pmatrix} 0 & \kappa \\ -\kappa & 0 \end{pmatrix} \begin{pmatrix} T \\ N \end{pmatrix}.$$

Example 60. $P(t) = \left(r\cos(t/r), r\sin(t/r)\right)$, $0 \le t \le 2\pi$. Then $P'(t) = \left(-\sin(t/r), \cos(t/r)\right)$. Since

$$\|P'(t)\|^2 = \sin^2(t/r) + \cos^2(t/r) = 1$$

P is a unit speed parametrization. Hence $T(t) = \left(-\sin(t/r), \cos(t/r)\right)$ and $N(t) = \left(-\cos(t/r), -\sin(t/r)\right)$. Since $T'(t) = \left(-\dfrac{1}{r}\cos(t/r), -\dfrac{1}{r}\sin t/r\right)$ we have $\kappa(t) = \dfrac{1}{r}$. Hence the curvature of a circle of radius r is $1/r$.

Our geometrical interpretation of curvature in the previous chapter can be further developed to prove the following:

$\kappa(t) < 0 \iff \theta(t)$ is strictly decreasing,
$\iff$ the curve lies "below" the tangent line,
$\iff$ the centre of curvature and the normal lie on opposite sides of the curve (see Figure 16.1).

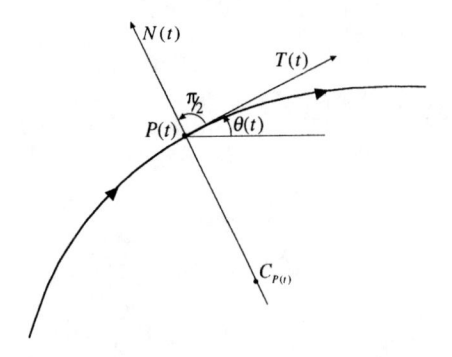

Figure 16.1

Vector Valued Differentiation

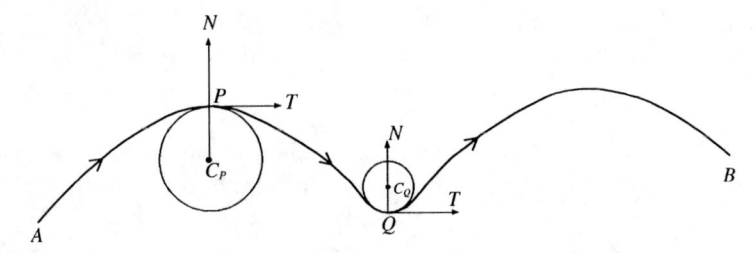

Figure 16.2

At the point P, in Figure 16.2, the circle of closest fit is on the opposite side of the curve to N and this denotes negative curvature, while at the point Q the circle of closest fit is on the same side of the curve as N and this denotes positive curvature.

In Figure 16.3 we see that the curve has positive curvature when it bends towards the normal and negative curvature when it bends away from the normal. At the points marked 0 the curvature is changing sign and takes the value 0. Not surprisingly, the curve is close to being a straight line at these points.

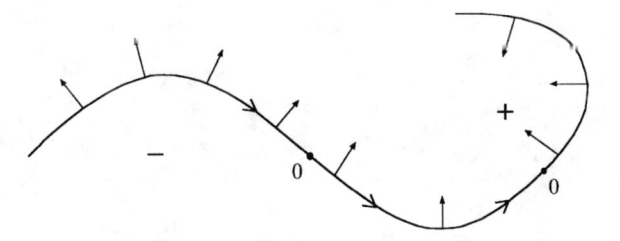

Figure 16.3

An ellipse will always have positive curvature since, **by definition**, it must be directed in an **anticlockwise** direction (Figure 16.4).

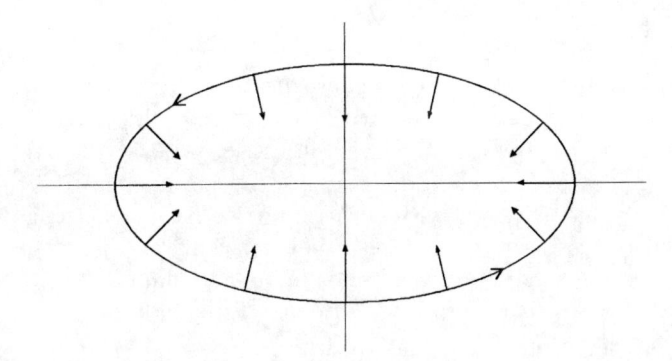

Figure 16.4

This completes our introduction to curves in the plane. This is the beginning of differential geometry, i.e., the study of geometrical objects using the differential calculus. It is an important part of modern mathematics and contains many results, even for plane curves, which have an immediate geometric appeal. We will quote one of these since we can readily appreciate it from our experience with the ellipse and since we have already encountered almost all the terms involved. A closed curve is **convex** if it always lies on **one** side of **each** tangent line. In Figure 16.5 we have a convex curve, while in Figure 16.6 the curve is not convex.

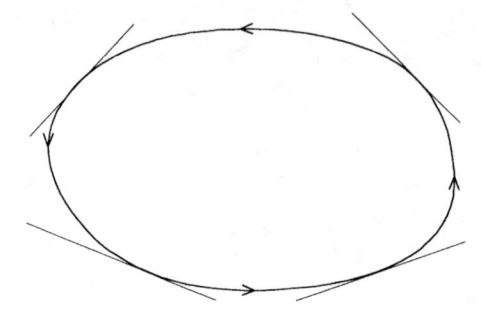

Figure 16.5

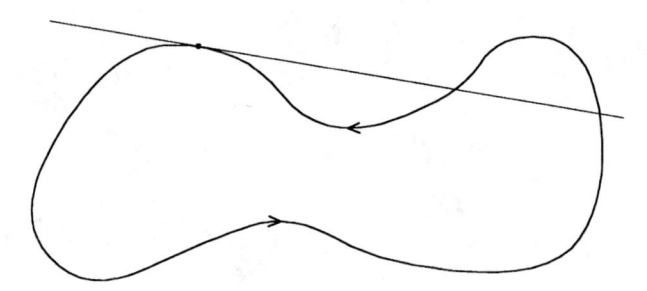

Figure 16.6

From our comments above it is not surprising that a closed curve is convex if and only if it has non-negative curvature. A slightly more specialized closed curve, **an oval**, is a curve with strictly positive curvature. Since all the curves we consider admit unit speed parametrization this implies we have excluded curves with corners but we have something similar, "a rounded corner", which is called a **vertex**. This is a point on the curve where the curvature has a local maximum or minimum. The **four vertex theorem** states that an oval has **at least four** vertices.

Exercises

16.1 If $f : \mathbf{R} \to \mathbf{R}^2$ is a differentiable function and $f \neq 0$ show that $\|f\|$ is differentiable. By considering the function $f(t) = (t, 0)$ show that the non-zero hypothesis is sometimes necessary. Is this always the case? What about $f(t) = (t^2, 0)$? Is $\|f\|$ differentiable at the origin? If $f \neq 0$ show that
$$\frac{d}{dt}(\|f\|) = \frac{\langle f', f \rangle}{\|f\|}.$$

16.2 Find $\dfrac{d}{dt}(\|(\cos^3 t, \sin^3 t)\|)$.

16.3 Verify the formula $\nabla(fg) = f\nabla g + g\nabla f$ for $f(x,y) = x^2 - y^2$, $g(x,y) = x^2 + y^2$.

16.4 Find the vertices on the level set $18x^2 + 25y^2 = 1$.

16.5 A set U in $\mathbf{R}^2$ is convex if for any pair of points a, b in U the straight line segment joining a and b lies in U. Show that the intersection of any number of convex sets is convex. Using this result show that the inside of a closed convex curve in $\mathbf{R}^2$ is a convex open subset of $\mathbf{R}^2$.

16.6 Let $f : U \to \mathbf{R}$ have continuous first and second order partial derivatives at all points in the convex open subset U of $\mathbf{R}^2$. If (a,b) is a critical point of f show, using Exercise 1.11 or otherwise, that f has a maximum at (a,b) if $f_{xx}(a,b) \leq 0$, $f_{yy}(a,b) \leq 0$ and $\det(H_{f(x,y)}) \geq 0$ for all $(x,y) \in U$. Using this result show that $f(x,y) = x + y - e^x - e^y - e^{x+y}$ attains its maximum over $\mathbf{R}^2$ when $x = y = \log((-1 + \sqrt{5})/2)$.

17

Complex Analysis

Summary. *We identify $\mathbf{R}^2$ with the complex plane $\mathbf{C}$ and define differentiable functions of a complex variable. We obtain the Cauchy-Riemann equations and show that the real and imaginary parts of a holomorphic function are harmonic. A model for heat transfer using harmonic and holomorphic functions is briefly discussed.*

We have considered functions from $\mathbf{R}$ into $\mathbf{R}^2$ and from $\mathbf{R}^2$ into $\mathbf{R}$ and we now consider the final set of functions, functions from $\mathbf{R}^2$ into $\mathbf{R}^2$. We can treat such functions as a pair of functions from $\mathbf{R}^2$ into $\mathbf{R}$ and apply the methods we have developed. This, however, neglects other structures that are available to us on $\mathbf{R}^2$. In particular, by identifying $\mathbf{R}^2$ and $\mathbf{C}$ we obtain a new theory.

Let $i = \sqrt{-1}$. The identification of (x,y) with $z = x + iy$ establishes a correspondence between $\mathbf{R}^2$ and $\mathbf{C}$. Addition of vectors corresponds to addition of complex numbers and multiplication of a vector by a real number corresponds to multiplication of a complex number by a real number. However, for complex numbers we have one further operation which we do not have for vectors—**multiplication** of complex numbers.

We consider a function f from an open set Ω in $\mathbf{C}$ (which is the same as an open set in $\mathbf{R}^2$) into $\mathbf{C}$. We let z denote the variable in Ω. Then $f(z)$ is a complex number and so has a **real** and **imaginary** part which we denote by $u(z)$ and $v(z)$, respectively. With this notation

$$f(z) = u(z) + iv(z)$$

and, for simplicity, we write

$$w = f(z) \quad \text{and} \quad f = u + iv.$$

Complex Analysis

We say that f is **C-differentiable** or **holomorphic** at z if
$$\lim_{\Delta z \to 0} \frac{f(z+\Delta z) - f(z)}{\Delta z}$$
exists and we denote this limit by $f'(z)$ or by $\dfrac{df}{dz}$. This is exactly the same situation as in the **one real variable** case but now we are dealing with **one complex variable**.

As in the real case we see—with the same proof we gave earlier—that f is **C**-differentiable at z if and only if
$$f(z+\Delta z) = f(z) + f'(z) \cdot \Delta z + g(z, \Delta z) \cdot \Delta z$$
where $g(z, \Delta z) \to 0$ as $\Delta z \to 0$. Moreover, if
$$f(z+\Delta z) = \alpha + \beta \Delta z + h(z, \Delta z) \cdot \Delta z$$
and $h(z, \Delta z) \to 0$ as $\Delta z \to 0$, then $\alpha = f(z), \beta = f'(z)$ and $h(z, \Delta z) = g(z, \Delta z)$.

Let $f'(z) = A + iB$ where A and B are real numbers. Using the identification of **C** with $\mathbf{R}^2$, we see that u and v can be regarded as two real-valued functions from an open subset of $\mathbf{R}^2$ into $\mathbf{R}$. When we regard them in this fashion we write $u(x, y)$ and $v(x, y)$ in place of $u(z)$ and $v(z)$. Rewriting the formula for $f(z+\Delta z)$ in this way and letting
$$\Delta z = \Delta x + i\Delta y \text{ and } g = g_1 + ig_2$$
we have
$$u(x+\Delta x, y+\Delta y) + iv(x+\Delta x, y+\Delta y)$$
$$= u(x,y) + iv(x,y) + (A+iB) \cdot (\Delta x + i\Delta y)$$
$$+ (g_1, -g_2) \cdot (\Delta x, \Delta y) + i(g_2, g_1) \cdot (\Delta x, \Delta y).$$

If we equate the real and imaginary parts of both sides we get
$$u(x+\Delta x, y+\Delta y) = u(x,y) + A\Delta x - B\Delta y + (g_1, -g_2) \cdot (\Delta x, \Delta y)$$
$$v(x+\Delta x, y+\Delta y) = v(x,y) + A\Delta y + B\Delta x + (g_2, g_1) \cdot (\Delta x, \Delta y).$$

Since, for fixed x and y, both g_1 and g_2 tend to 0 as Δx and Δy tend to 0, u and v are differentiable at (x, y) and, moreover,
$$\nabla u = (A, -B) \text{ and } \nabla v = (B, A).$$

This has some amazing consequences. To begin with, since $\nabla u = \left(\dfrac{\partial u}{\partial x}, \dfrac{\partial u}{\partial y}\right)$ and $\nabla v = \left(\dfrac{\partial v}{\partial x}, \dfrac{\partial v}{\partial y}\right)$, we have
$$A = \frac{\partial u}{\partial x}, \quad B = -\frac{\partial u}{\partial y}, \quad B = \frac{\partial v}{\partial x} \text{ and } A = \frac{\partial v}{\partial y}.$$

From this we get two equations—the **Cauchy-Riemann equations**—
$$\frac{\partial u}{\partial x} = \frac{\partial v}{\partial y} \quad \text{and} \quad \frac{\partial u}{\partial y} = -\frac{\partial v}{\partial x}.$$
These equations are satisfied by the real and imaginary parts of every **C**-differentiable function, and the converse can also be proved. In other words, if we are given two functions from $\mathbf{R}^2 \to \mathbf{R}$ with continuous first order partial derivatives, then they can be combined to form a **C**-differentiable function if and only if they satisfy the Cauchy-Riemann equations. We have
$$f'(z) = A + iB = \frac{\partial u}{\partial x} - i\frac{\partial u}{\partial y}$$
$$= i\frac{\partial v}{\partial x} + \frac{\partial v}{\partial y}.$$
So if we know either u or v then we can calculate f'.

Example 61. Let $f(z) = z^3$. By the binomial theorem
$$\begin{aligned} f(z) = (x + iy)^3 &= x^3 + 3x^2(iy) + 3x(iy)^2 + (iy)^3 \\ &= x^3 + 3ix^2y - 3xy^2 - iy^3 \\ &= x^3 - 3xy^2 + i(3x^2y - y^3). \end{aligned}$$
So $u(x, y) = x^3 - 3xy^2$ and $v(x, y) = 3x^2y - y^3$. Hence
$$\frac{\partial u}{\partial x} = 3x^2 - 3y^2, \quad \frac{\partial v}{\partial y} = 3x^2 - 3y^2$$
and
$$\frac{\partial u}{\partial x} = \frac{\partial v}{\partial y}.$$
We also have
$$\frac{\partial u}{\partial y} = -6xy \quad \text{and} \quad \frac{\partial v}{\partial x} = 6xy$$
and
$$\frac{\partial u}{\partial y} = -\frac{\partial v}{\partial x}.$$
Moreover,
$$\begin{aligned} f'(z) = \frac{\partial u}{\partial x} - i\frac{\partial u}{\partial y} &= 3x^2 - 3y^2 + i6xy \\ &= 3(x^2 - y^2 + 2ixy) \\ &= 3z^2. \end{aligned}$$

Example 62. If $v(x, y) = 0$ for all x and y and f is **C**-differentiable then f is a constant function. We have $\dfrac{\partial v}{\partial x} = \dfrac{\partial v}{\partial y} = 0$ and hence $f'(z) = i\dfrac{\partial v}{\partial x} + \dfrac{\partial v}{\partial y} = 0$ and it is not too difficult to show this implies that f is a constant function. Alternatively, by the Cauchy-Riemann equations we get $\dfrac{\partial u}{\partial x} = \dfrac{\partial u}{\partial y} = 0$. From the one variable theory we see that u does not depend on either x or y and so is a constant function. It was because of this example that we did not define in Chapter 1 a derivative of functions from $\mathbf{R}^2 \to \mathbf{R}$ by identifying $\mathbf{R}^2$ with $\mathbf{C}$. The only differentiable functions with this definition are the constant functions.

We complete this brief introduction to **C**-differentiable functions with two more simple results.

A function from $\mathbf{R}^2$ into $\mathbf{R}$ with continuous second order partial derivatives is called **harmonic** if it satisfies the following partial differential equation, **Laplace's equation**:

$$\frac{\partial^2 u}{\partial x^2} + \frac{\partial^2 u}{\partial y^2} = 0.$$

Proposition 63. *The real and imaginary parts of **C**-differentiable functions are harmonic.*

Proof. To prove this we must assume that the real and imaginary parts have continuous second order partial derivatives (this is in fact true of all **C**-differentiable functions). We prove the result only for the real part u; the imaginary part is handled in the same way. We have, by the Cauchy-Riemann equations,

$$\begin{aligned}\frac{\partial^2 u}{\partial x^2} + \frac{\partial^2 u}{\partial y^2} &= \frac{\partial}{\partial x}\left(\frac{\partial u}{\partial x}\right) + \frac{\partial}{\partial y}\left(\frac{\partial u}{\partial y}\right) \\ &= \frac{\partial}{\partial x}\left(\frac{\partial v}{\partial y}\right) + \frac{\partial}{\partial y}\left(-\frac{\partial v}{\partial x}\right), \\ &= \frac{\partial^2 v}{\partial x \partial y} - \frac{\partial^2 v}{\partial y \partial x} = 0.\end{aligned}$$

For our final result we return to the formulae we found for ∇u and ∇v. We showed $\nabla u = (A, -B)$ and $\nabla v = (B, A)$. If we take the dot product we get

$$\nabla u \cdot \nabla v = AB - BA = 0.$$

Hence the vectors ∇u and ∇v are perpendicular to one another. Now take a point z_0 where $f'(z_0) \neq 0$. Since $f'(z_0) = A + iB$ and $\nabla u = (A, -B)$

and $\nabla v = (B, A)$ it follows that $\nabla u(x_0, y_0)$ and $\nabla v(x_0, y_0)$ are non-zero. Let $f(z_0) = \alpha + i\beta$. If $u(x_0, y_0) = \alpha$ and $v(x_0, y_0) = \beta$ then the level sets $u(x, y) = \alpha$ and $v(x, y) = \beta$ pass through (x_0, y_0). Since $\nabla u(x_0, y_0)$ and $\nabla v(x_0, y_0)$ are both non-zero it follows that both level sets have tangents at (x_0, y_0) and also both have normals. Now the normal to the level set $u(x, y) = \alpha$ at (x_0, y_0) is $\nabla u(x_0, y_0)$ and the normal to the level set $v(x, y) = \beta$ is $\nabla v(x_0, y_0)$. Since $\nabla u(x_0, y_0) \cdot \nabla v(x_0, y_0) = 0$ it follows that the normals are perpendicular to one another. Hence the tangents are also perpendicular to one another.

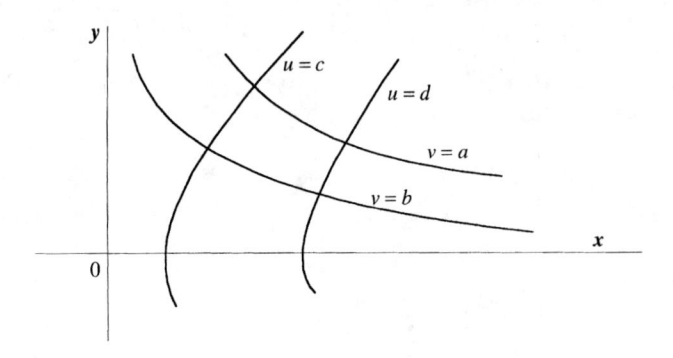

Figure 17.1

If we define the **angle between two curves** at a point of intersection as the angle between their tangents at the point of intersection, then we have proved the following.

Proposition 64. *Level sets of the real and imaginary parts of a C-differentiable function intersect at right angles (i.e., are orthogonal) when the C-derivative is non-zero.*

This proposition can be illustrated by considering even the most elementary functions. For example, if $f(z) = z^2$ then $u(x, y) = x^2 - y^2$ and $v(x, y) = 2xy$. Thus the level sets of both u and v are hyperbolas (see Figure 17.1). At all points, except the origin, these intersect orthogonally.

Complex analysis is an important area of pure mathematics with useful applications in physics and electronic engineering. We briefly describe a simple model of heat transfer in a flat (hence two-dimensional) conducting surface with insulated boundary apart from a **source** (of heat) and a **sink** (which absorbs heat). We assume that conditions are ideal so that we are in a **steady state** situation (i.e., the **temperature** $T(x, y)$ at any point does not depend on time). We assume the reasonable physical law that we experience ourselves every day, namely that **heat** (i.e., **thermal energy**) transfers itself as efficiently as possible from hotter to colder regions. If we take a level set of T, an **isotherm**, then clearly there will be no transfer of heat along such a set. By Example 40, the direction of greatest change of T will be in the direction $\nabla T = (T_x, T_y)$. The function ∇T is called the **heat flux density** as it is, by the above, related to the intensity of heat conduction. The steady state condition can be used to prove that

$$\frac{\partial}{\partial x}(T_x) + \frac{\partial}{\partial y}(T_y) = 0 \, ,$$

i.e.,

$$\frac{\partial^2 T}{\partial x^2} + \frac{\partial^2 T}{\partial y^2} = 0$$

and thus the temperature is a harmonic function. Now that T is harmonic it can be shown that there exists a holomorphic function T_c, the **complex temperature**, such that T is the real part of T_c.

The imaginary part of T_c, S, is called the **stream function** in applied mathematics and the **harmonic conjugate** in pure mathematics. This terminology comes about as follows. By Proposition 64 the level sets of S, the **stream lines**, are perpendicular to the isotherms and hence the tangents to the stream lines are parallel to ∇T and give the direction in which heat appears to flow (as in a stream).

It is remarkable that the **same** model applies to fluid flow and electrostatics. A small dictionary shows how to go from one system to another.

Exercises

17.1 If c and d are positive real numbers and $d > 2c$ show that

$$\{z; |z-c| + |z+c| = d\} = \{(x,y); \frac{4x^2}{d^2} + \frac{4y^2}{d^2 - 4c^2} = 1\}.$$

17.2 Find the real and imaginary parts of the functions e^z, $\sinh z$, $\cosh z$, ze^z, $z \cos z$. Prove that the real and imaginary parts of each of these functions satisfies the Cauchy-Riemann equations and that they are harmonic (see also Exercise 3.5).

17.3 Let $f(z) = ze^z$. Find $f'(z)$ using only the real part of f. Verify that $\nabla u \cdot \nabla v = 0$ where u and v are the real and imaginary parts of f.

17.4 By examining the function $f(z) = z^2$ at $z = 0$ show that the condition $f'(z) \neq 0$ is necessary in Proposition 64.

17.5 Let $f(z) = \frac{1}{2}(z + z^{-1})$. Find the real and imaginary parts of f. Show that f maps the circle $|z| = r$, $0 < r < 1$, onto the ellipse

$$\frac{4r^2 u^2}{(1+r^2)^2} + \frac{4r^2 v^2}{(1-r^2)^2} = 1.$$

18

Line Integrals

Summary. We define line integrals of real-valued functions along a directed curve. We show that such integrals can be expressed as integrals over intervals in **R**. Examples are given.

The remainder of this book is devoted to integration theory. We discuss integrals along a curve (often called **line integrals**) and integrals over the interior of regions bounded by a curve (called **double integrals**). We begin by recalling how integrals of functions of one variable over an interval $[a,b]$ are defined. This approach to integration is of little use in evaluating integrals but understanding it is extremely helpful as a guide to the higher dimensional theory.

So how do we define $\int_a^b f(x)\,dx$?

We start by partitioning the interval $[a,b]$ into subintervals (Figure 18.1).

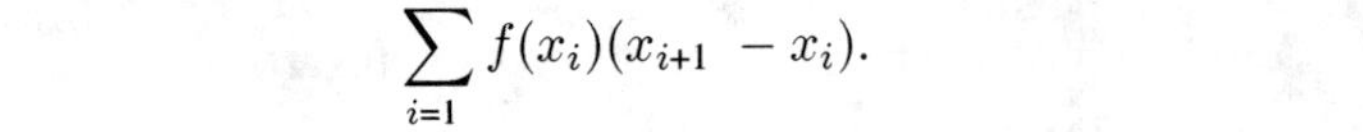

$$a = x_0 \quad x_1 \quad \cdots \quad x_i \quad x_{i+1} \quad \cdots \quad x_n = b$$

Figure 18.1

With the **partition** $P = [x_0, x_1, \ldots, x_n]$ of the domain we form the **Riemann sum**

$$\sum_{i=1}^{n} f(x_i)(x_{i+1} - x_i).$$

of the function f. If, as we take finer and finer partitions, we find that this sum tends to a limit, then we say that f is **integrable** (or **Riemann integrable**) and we call the limit the integral of f between a and b. We write the limit as

$$\int_a^b f(x)\,dx.$$

Line Integrals

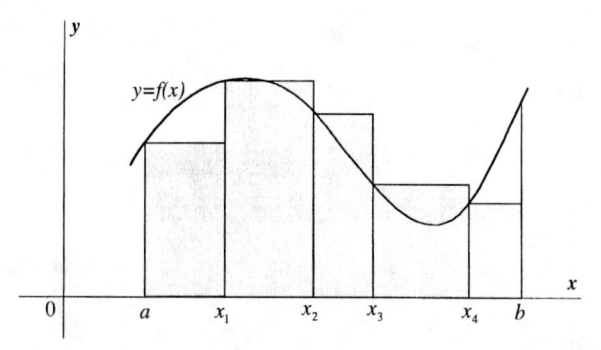

Figure 18.2

The Riemann sum can be represented diagrammatically as shown in Figure 18.2. The shaded portion is a Riemann sum. If f is a positive function and $\int_a^b f(x)\,dx$ exists, then

$$\int_a^b f(x)\,dx = \text{Area under the graph of } f \text{ between } a \text{ and } b.$$

Another important interpretation of the integral occurs in probability theory. This is as the **expected value** or **average of a random variable**. This definition of integral will only be useful if sufficiently many interesting functions are integrable. It can be shown that any **continuous** function on a closed interval of finite length is integrable. To prove this one needs the concept of **uniform continuity** and also, since the collection of all partitions of an interval do not form a sequence, the idea of limit over a "**generalized sequence**" or **net**. We shall not discuss these ideas any further other than to remark that the same concepts are also the key ingredients in a rigorous treatment of all the integrals we discuss. In the definition of the Riemann sum we used a partition of the domain of f. Replacing this by a partition of the range of f leads to another kind of integral—the **Lebesgue integral**—which is also important in modern mathematics and probability theory.

The Riemann sum is often written in the form

$$\sum_{i=1}^{n} f(x_i)\Delta x_i$$

where $\Delta x_i = x_{i+1} - x_i$ and in this form it is readily compared with the limit

$$\int_a^b f(x)\,dx.$$

The symbol $\int$ is a corruption of the symbol $\sum$, the Greek letter sigma, which is the equivalent of S, and the letter S is used to denote sum.

140   *Functions of Two Variables*

In the integrals we discuss, it is relatively simple to write down the Riemann sum and, using the comparison of $\sum_{i=1}^{n} f(x_i) \Delta x_i$ with $\int_a^b f(x)\, dx$ as a guide, it is then easy to see what meaning we should give to our integrals.

We begin by looking at $\int_\Gamma F\, dx$ where F is a real-valued **continuous** function of two variables and Γ is the graph of the **continuous** function $y = f(x)$ over the interval $[a, b]$ (Figure 18.3).

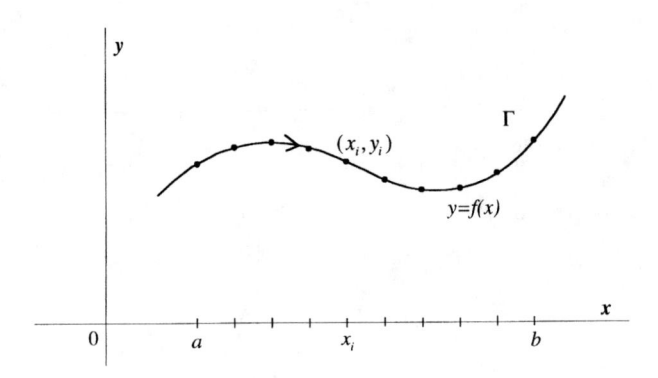

Figure 18.3

We again start with a partition $P = [x_0, x_1, \ldots, x_n]$ of $[a, b]$ and form the Riemann sum

$$\sum_{i=1}^{n} F(x_i, y_i) \cdot (x_{i+1} - x_i)$$

where y_i is the point on the graph of $y = f(x)$ above x_i, in other words $y_i = f(x_i)$. Hence the Riemann sum can be rewritten as

$$\sum_{i=1}^{n} F(x_i, f(x_i)) \cdot \Delta x_i$$

and if the limit exists then the one-dimensional theory shows that the limit is

$$\int_a^b F(x, f(x))\, dx.$$

Next we consider $\int_\Gamma F\, dy$. To apply the same method we require the curve Γ to be the graph of some function of y, say $x = g(y)$ (in fact $g = f^{-1}$). This will be the case if each horizontal line cuts the graph in precisely one

Line Integrals

point (Figure 18.4). When this is the case we form a partition on the y-axis. The corresponding Riemann sum is

$$\sum_{j=1}^{m} F(x_j, y_j) \cdot (y_{j+1} - y_j) = \sum_{j=1}^{m} F(g(y_j), y_j) \cdot \Delta y_j$$

and the limit is $\int_c^d F(g(y), y)\, dy$.

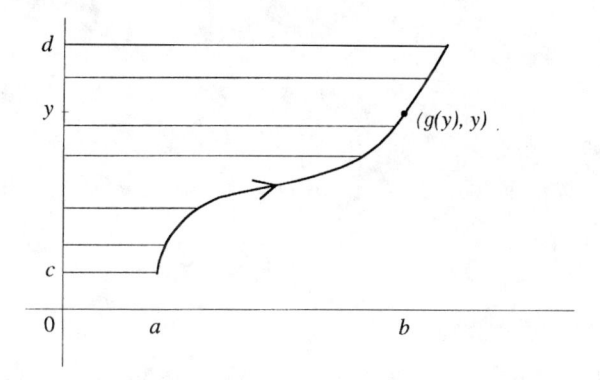

Figure 18.4

Example 65. Let Γ denote the graph of the function $y = x^2$ between 0 and 2, and let $F(x, y) = x^2 y$. We wish to find $\int_\Gamma F\, dx$ and $\int_\Gamma F\, dy$.

In the case of $\int_\Gamma F\, dx$ we have $y = x^2$ and x varies between 0 and 2. Hence

$$\int_\Gamma F\, dx = \int_0^2 F(x, y)\, dx = \int_0^2 F(x, x^2)\, dx$$
$$= \int_0^2 x^2 \cdot x^2\, dx = \left. \frac{x^5}{5} \right|_0^2 = 32/5.$$

We are told in this case that we are integrating along the graph of a function of x and so were able to proceed directly. To evaluate $\int_\Gamma F\, dy$ we have to check that we are integrating along a graph of a function of y.

By inspection this is the graph of a function of y since each horizontal line cuts the graph only once. What is the function? Well, if $y = x^2$ then

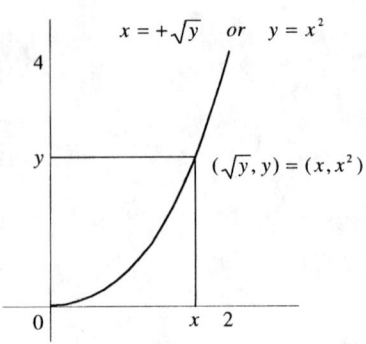

Figure 18.5

$x = \sqrt{y}$ (Figure 18.5). Hence

$$\int_1 F \, dy = \int_0^4 F(x,y) \, dy = \int_0^4 F(\sqrt{y}, y) \, dy$$
$$= \int_0^4 y \cdot y \, dy = \int_0^4 y^2 \, dy = \left.\frac{y^3}{3}\right]_0^4 = 64/3.$$

Notice that we used the sketch to find the limits of integration with respect to y, that the methods of integration used are the familiar ones from the one-dimensional theory—and this will always be the case—and that before integrating it was necessary to express everything to be integrated in terms of a single variable—x in the case of $\int F \, dx$ and y in the case of $\int F \, dy$.

Next we consider the problem of integrating over an arbitrary directed curve Γ. Let $P : [a, b] \to \mathbf{R}^2$ denote a parametrization of Γ. Suppose $P(t) = (x(t), y(t))$. Let F denote a continuous real-valued function of two variables which is defined at all points of the set $P([a, b]) = \Gamma$.

We partition Γ as in Figure 18.6. If we denote a typical point by (x_i, y_i)

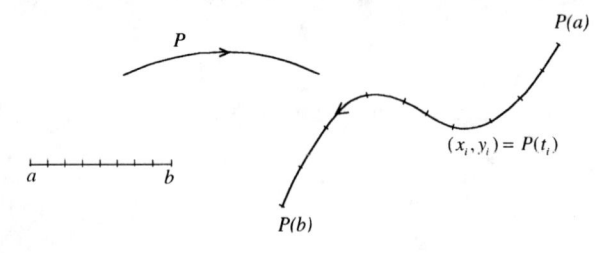

Figure 18.6

Line Integrals

we see that, for each i, there is a t_i in $[a, b]$ such that $(x_i, y_i) = P(t_i)$ Thus, any partition of Γ leads to a partition of $[a, b]$.

If we are interested in $\int_\Gamma f\, dx$ we form the Riemann sum

$$\sum_{i=1}^n f(x_i, y_i) \cdot \Delta x_i$$

which is equal to

$$\sum_{i=1}^n f\bigl(x(t_i), y(t_i)\bigr) \cdot \bigl(x(t_{i+1}) - x(t_i)\bigr)$$

$$= \sum_{i=1}^n f\bigl(x(t_i), y(t_i)\bigr) \frac{x(t_{i+1}) - x(t_i)}{t_{i+1} - t_i} \Delta t_i$$

and taking the limit we get

$$\int_\Gamma f(x, y)\, dx = \int_a^b f\bigl(x(t), y(t)\bigr) x'(t)\, dt. \tag{$*$}$$

Similarly

$$\int_\Gamma f(x, y)\, dy = \int_a^b f\bigl(x(t), y(t)\bigr) y'(t)\, dt. \tag{$**$}$$

To remember this, note that $x = x(t)$ so $\dfrac{dx}{dt} = x'(t)$ and $dx = x'(t)\, dt$. Similarly $dy = y'(t)\, dt$. Thus in $(*)$ and $(**)$ replace x by $x(t)$, y by $y(t)$, dx by $x'(t)dt$ and dy by $y'(t)dt$. This gives an integral in the single variable t over $[a, b]$.

We now introduce a new compact notation for integrals of the form

$$\int_\Gamma f(x, y)\, dx \quad \text{and} \quad \int_\Gamma g(x, y)\, dy$$

where Γ is a directed curve. In place of considering scalar-valued functions we consider

$$F = (f, g) : U \subset \mathbf{R}^2 \longrightarrow \mathbf{R}^2$$

and we suppose that the domain of definition of U contains Γ. We suppose F is continuous (or equivalently that f and g are both continuous). Let $P(t) : [a, b] \longrightarrow \Gamma$ denote a parametrization of Γ. We define $\int_\Gamma F$ to be

$$\int_a^b F(P(t)) \cdot P'(t)\, dt.$$

Note that $F(P(t)) \in \mathbf{R}^2$ and $P'(t) \in \mathbf{R}^2$ and the use of the dot product implies $F(P(t)) \cdot P'(t) \in \mathbf{R}$. Since F and P' are both continuous, the function $t \in [a,b] \longrightarrow F(P(t)) \cdot P'(t)$ is a continuous real-valued function.

If $P(t) = (x(t), y(t))$, then $P'(t) = (x'(t), y'(t))$. We have

$$\int_\Gamma f(x,y)\,dx = \int_a^b f(x(t), y(t)) x'(t)\,dt$$

and

$$\int_\Gamma g(x,y)\,dy = \int_a^b g(x(t), y(t)) y'(t)\,dt$$

for any pair of continuous functions f and g defined on a subset of $\mathbf{R}^2$ containing Γ. We add these together and, using the dot product $\langle\,,\,\rangle$, we get

$$\int_\Gamma f(x,y)\,dx + \int_\Gamma g(x,y)\,dy$$
$$= \int_a^b f(x(t), y(t)) x'(t)\,dt + \int_a^b g(x(t), y(t)) y'(t)\,dt$$
$$= \int_a^b \left\langle \big(f(x(t), y(t)), g(x(t), y(t)) \big), (x'(t), y'(t)) \right\rangle dt$$
$$= \int_a^b \left\langle \big(f(P(t)), g(P(t)) \big), P'(t) \right\rangle dt$$
$$= \int_a^b F(P(t)) \cdot P'(t)\,dt.$$

We have shown that

$$\int_\Gamma F = \int_a^b F(P(t)) \cdot P'(t)\,dt = \int_\Gamma f(x,y)\,dx + g(x,y)\,dy.$$

This gives us a further method for evaluating line integrals over directed curves which, in certain circumstances, depending on the form of F and P, may lead to simpler calculations.

We can also rewrite the above using the tangent since $T(t) = \dfrac{P'(t)}{\|P'(t)\|}$.

We have

$$\int_a^b F(P(t)) \cdot P'(t)\,dt = \int_a^b F(P(t)) \cdot \frac{P'(t)}{\|P'(t)\|} \cdot \|P'(t)\|\,dt$$
$$= \int_a^b (F \cdot T)\,ds.$$

To derive the final integral we recall, from Chapter 13, that $s(t)$ is the distance travelled up to time t. Hence $s'(t) = \|P'(t)\|$ is the speed and

Line Integrals

$ds = \|P'(t)\|\, dt$. The unit tangent is denoted by T and in evaluating this integral the function $F \cdot T$ is evaluated at $P(t)$.

It is important not to forget that a line integral is always over a **directed curve**. In our first two definitions of a line integral over a graph, we did not explicitly mention this fact but an examination of our proof in, for instance, the case $\int_\Gamma F\, dx$ shows that we used implicitly the parametrization $x \to (x, f(x))$ and this makes the graph of f a directed curve. This, after all, is also what we do in the one variable case when we distinguish between

$$\int_a^b f(x)\, dx \quad \text{and} \quad \int_b^a f(x)\, dx.$$

As in the one variable case changing the direction changes the sign of the integral. If $\Gamma \equiv (\Gamma, A, B)$ is a directed curve, $A \neq B$, and $\tilde{\Gamma} = (\Gamma, B, A)$, see Figure 13.1, then

$$\int_{\tilde{\Gamma}} = -\int_\Gamma.$$

Example 65. Find $\int_\Gamma x^3 y^2\, dx + x^2 y^3\, dy$ where Γ is the part of the ellipse $\dfrac{x^2}{4} + \dfrac{y^2}{9} = 1$ above the x-axis (Figure 18.7) with initial point $(2,0)$ and final point $(-2,0)$.

Let $P(t) = (2\cos t, 3\sin t)$. Since $P(0) = (2,0)$ and $P(\pi) = (-2,0)$ this is a parametrization of the directed curve Γ.

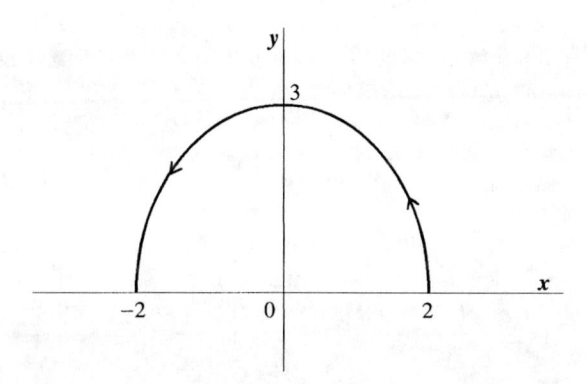

Figure 18.7

Let $F(x,y) = (x^3 y^2, x^2 y^3)$. We have

$$F(P(t)) = F(2\cos t, 3\sin t) = (72\cos^3 t \sin^2 t, 108\cos^2 t \sin^3 t),$$

and
$$P'(t) = (-2\sin t, 3\cos t).$$

Hence
$$F(P(t)) \cdot P'(t) = -144\cos^3 t \sin^3 t + 324 \cos^3 t \sin^3 t$$
$$= 180 \cos^3 t \sin^3 t$$

and
$$\begin{aligned}
\int_1 x^3 y\, dx + x^2 y^3\, dy &= \int_1 F \\
&= \int_0^\pi F(P(t)) \cdot P'(t)\, dt \\
&= 180 \int_0^\pi \cos^3 t \sin^3 t\, dt \\
&= \frac{45}{2} \int_0^\pi \sin^3 2t\, dt \\
&= \frac{45}{2} \int_0^\pi (1 - \cos^2 2t) \sin 2t\, dt \\
&= \frac{45}{2} \int (1 - u^2) \frac{du}{-2} \quad \begin{array}{l} \text{Letting } u = \cos 2t \\ du = -2\sin 2t\, dt \end{array} \\
&= -\frac{45}{4}\left[u - \frac{u^3}{3}\right] \\
&= -\frac{45}{4}\left[\cos 2t - \frac{\cos^3 2t}{3}\right]_0^\pi \\
&= 0.
\end{aligned}$$

We have defined line integrals of continuous functions over graphs of continuous functions with continuous inverses and over curves, parametrized by differentiable functions, using different methods. Neither situation contains the other although both methods can be used for many examples. We now show, **when both methods apply**, that they lead to the same value of the integral.

If f is a continuously differentiable function on $[a, b]$ and $f'(t) \neq 0$ for all $t \in [a, b]$ then f^{-1} exists and is defined on the interval $[f(a), f(b)]$ (note that we may have $f(a) > f(b)$). The graph of f, Γ, is also the graph of f^{-1}. If G and H are functions of two variables, the definition of **integral over a graph**, implies

$$\int_\Gamma G\, dx + H\, dy = \int_a^b G(x, f(x))\, dx + \int_{f(a)}^{f(b)} H(f^{-1}(y), y)\, dy.$$

If we parameterize Γ by $P(t) = (t, f(t))$, $a \leq t \leq b$, (see Example 56) then,

Line Integrals

using the definition of **integral** over **a parametrized curve**, we obtain

$$\int_\Gamma G\,dx + H\,dy = \int_a^b G(t, f(t))\,dt + \int_a^b H(t, f(t))f'(t)\,dt.$$

The substitution $y = f(t)$, $dy = f'(t)\,dt$ and $t = f^{-1}(y)$, shows that

$$\int_a^b H(t, f(t))f'(t)\,dt = \int_{f(a)}^{f(b)} H(f^{-1}(y), y)\,dy$$

and both definitions agree when we **direct** the graph Γ from **left to right**.

Exercises

18.1 Evaluate $\int_\Gamma xy^2\,dx$ and $\int_\Gamma xy^2\,dy$, along the part of the unit circle joining $(1, 0)$ to $(0, -1)$ in an anticlockwise fashion.

18.2 Evaluate $\int_\Gamma f(x, y)\,dx$, $\int_\Gamma f(x, y)\,dy$ where

    (a) $f(x, y) = x^3 + y$, $x = 3t$, $y = t^3$, $0 \le t \le 1$,

    (b) $f(x, y) = xy^{2/5}$, $x = t/2$, $y = t^{5/2}$, $0 \le t \le 1$.

    (c) $f(x, y) = (x/y)\sin(y/x)$, $x = t^2, y = t^3, 1/2 \le t \le 1$.

18.3 Evaluate $\int_\Gamma xe^{xy}\,dy$ where Γ is the graph of $y = x^3$ with initial point $(-1, -1)$ and final point $(1, 1)$.

Find $\int_\Gamma e^{x-y}\,dy$ where Γ is the straight line segment joining $(-1, 1)$ to $(1, 2)$.

18.4 Evaluate $\int_\Gamma xy\,dx + (x + y)\,dy$ where Γ is the anticlockwise oriented triangle with vertices $(0, 0), (2, 0), (1, 1)$.

19

The Fundamental Theorem of Calculus

Summary. *We obtain the fundamental theorem of calculus for line integrals. We introduce the concept of the potential of a function from* $\mathbf{R}^2 \to \mathbf{R}^2$, *show how to identify functions having a potential and how to find a potential.*

In our investigations we have used the following two versions of the fundamental theorem of calculus:

if f is differentiable then

$$\int_a^b f'(t)\,dt = f(b) - f(a); \tag{1}$$

if f is continuous and $a \leq x \leq b$ then

$$\frac{d}{dx} \int_a^x f(t)\,dt = f(x). \tag{2}$$

At first glance both are similar in showing that differentiation and integration are inverse operations. We have

$$\int \frac{d}{dt}(f) = \frac{d}{dx} \int f = f.$$

A second glance shows that (1) implies the values of certain integrals can be calculated from values on the boundary of the domain of integration, while (2) allows one to construct functions with prescribed derivatives. Thus (1) is the main tool used to calculate definite integrals in the single variable theory and we provide later, in the form of Green's theorem, a two-dimensional version of this result.

In this chapter we generalize (1) and (2) using line integrals. Not unexpectedly, the gradient plays the role of derivative in our results.

The Fundamental Theorem of Calculus

Theorem 66. *If $P : [a,b] \to \Gamma$ is a parametrized curve in $\mathbf{R}^2$, $A = P(a)$ and $B = P(b)$, and $f : \mathbf{R}^2 \to \mathbf{R}$ has continuous first order partial derivatives, then*
$$\int_\Gamma \nabla f = f(B) - f(A).$$

Proof. Let $g(t) = f \circ P(t)$ for t in $[a,b]$. By the chain rule
$$g'(t) = f'(P(t)) \cdot P'(t) = \nabla f(P(t)) \cdot P'(t).$$
By the fundamental theorem of calculus for one variable we have
$$\int_\Gamma \nabla f = \int_a^b \nabla f(P(t)) \cdot P'(t) \, dt$$
$$= \int_a^b g'(t) \, dt$$
$$= g(b) - g(a)$$
$$= f(P(b)) - f(P(a))$$
$$= f(A) - f(B).$$

We proceed to generalize (2), since the key is already contained in Theorem 66. We seek to identify those $F : \mathbf{R}^2 \to \mathbf{R}^2$ for which there exists $\phi : \mathbf{R}^2 \to \mathbf{R}$ such that $\nabla \phi = F$, i.e., given $F = (f,g)$ we solve the equations
$$\frac{\partial \phi}{\partial x} = f, \quad \frac{\partial \phi}{\partial y} = g.$$
When such a function ϕ exists it is called a **potential** of F. If F has a potential then Theorem 66 shows that $\int_\Gamma F$ will only depend on the end points of Γ and the next result shows that this is all we need.

Theorem 67. *Let U denote an open set in $\mathbf{R}^2$ and suppose every pair of points in U can be joined by a curve which lies in U (U is said to be* **connected** *in this case). Let $F = (f,g) : U \to \mathbf{R}^2$ denote a continuous function. If, for any two points A and B in U and any two directed curves Γ_1 and Γ_2 joining A and B, we have*
$$\int_{\Gamma_1} F = \int_{\Gamma_2} F$$
then there exists a differentiable function $\phi : U \to \mathbf{R}$ such that $\nabla \phi = F$.

Proof. Fix a point (x_0, y_0) in U and for any point (x,y) in U let Γ denote a parametrized curve joining (x_0, y_0) to (x,y). Let
$$\phi(x,y) = \int_\Gamma F.$$

By our hypothesis this does not depend on the Γ used and so ϕ is unambiguously defined. We must show $\dfrac{\partial \phi}{\partial x} = f$ and $\dfrac{\partial \phi}{\partial y} = g$. Since both are proved in the same way we just consider $\dfrac{\partial \phi}{\partial x}$. Fix (x, y) in U and choose Δx sufficiently small so that the straight line segment Γ_1 joining (x, y) to $(x + \Delta x, y)$ also lies in U. Let Γ_2 denote any curve in U joining (x_0, y_0) to (x, y) such that the tangent to Γ_2 at (x, y) is horizontal (see Figure 19.1).

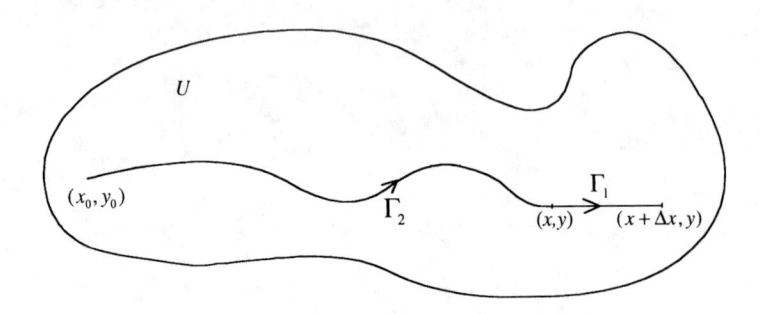

Figure 19.1

Since $\Gamma_1 \cup \Gamma_2$ is a curve joining (x_0, y_0) with $(x + \Delta x, y)$ we have

$$\phi(x + \Delta x, y) - \phi(x, y) = \int_{\Gamma_1 \cup \Gamma_2} F - \int_{\Gamma_2} F = \int_{\Gamma_1} F.$$

Now $P(t) = (x + t\Delta x, y)$, $0 \le t \le 1$, is a parametrization of Γ_1 and $P'(t) = (\Delta x, 0)$. Hence

$$\phi(x + \Delta x, y) - \phi(x, y)$$
$$= \int_0^1 F(P(t)) \cdot P'(t) \, dt$$
$$= \int_0^1 \left\langle \left(f(P(t)), g(P(t))\right), (\Delta x, 0) \right\rangle dt$$
$$= \Delta x \int_0^1 f(x + t\Delta x, y) \, dt$$
$$= f(x, y) \cdot \Delta x + \Delta x \int_0^1 \left(f(x + t\Delta x, y) - f(x, y)\right) dt.$$

Let $h(x, y, \Delta x) = \displaystyle\int_0^1 \left(f(x + t\Delta x, y) - f(x, y)\right) dt$. Since

$$f(x + \Delta x, y) - f(x, y) \to 0 \quad \text{as } \Delta x \to 0$$

it follows that

$$|h(x,y,\Delta x)| = \left|\int_0^1 (f(x+t\Delta x, y) - f(x,y))\, dt\right|$$
$$\leq \max_{|t|\leq |\Delta x|} |f(x+t, y) - f(x, y)| \to 0$$

as $\Delta x \to 0$.

We have shown that

$$\phi(x+\Delta x, y) = \phi(x, y) + f(x, y) \cdot \Delta x + h(x, y, \Delta x) \cdot \Delta x$$

where $h(x, y, \Delta x) \to 0$ as $\Delta x \to 0$. If $k(t) = \phi(x+t, y)$ then the above shows that

$$k(\Delta t) = k(0) + f(x, y)\Delta t + k_1(\Delta t) \cdot \Delta t$$

where $k_1(\Delta t) \to 0$ as $\Delta t \to 0$. Hence $k'(0) = f(x, y)$. Since $k'(0) = \dfrac{\partial \phi}{\partial x}(x, y)$ this proves that $\dfrac{\partial \phi}{\partial x} = f$. Similarly $\dfrac{\partial \phi}{\partial y} = g$. Since f and g are both continuous, ϕ has continuous first order partial derivatives and hence is differentiable. We have

$$\nabla \phi(x, y) = \left(\frac{\partial \phi}{\partial x}(x, y), \frac{\partial \phi}{\partial x}(x, y)\right) = (f(x, y), g(x, y)) = F$$

and this completes the proof.

Example 68. Consider the function $f : \mathbf{R}^2 \to \mathbf{R}^2$ given by $F(x, y) = (2xy, x^2)$. We consider two paths joining the points $(2, 0)$ and $(-2, 0)$, shown in Figure 19.2. Let Γ_1 denote the upper half of the circle $x^2 + y^2 = 4$ and let Γ_2 denote the lower half of the ellipse

$$\frac{x^2}{4} + y^2 = 1.$$

A parametrization of Γ_1 is given by $P_1(t) = (2\cos t, 2\sin t)$, $0 \leq t \leq \pi$, and a parametrization of Γ_2 by $P_2(t) = (2\cos t, -\sin t)$, $0 \leq t \leq \pi$. We

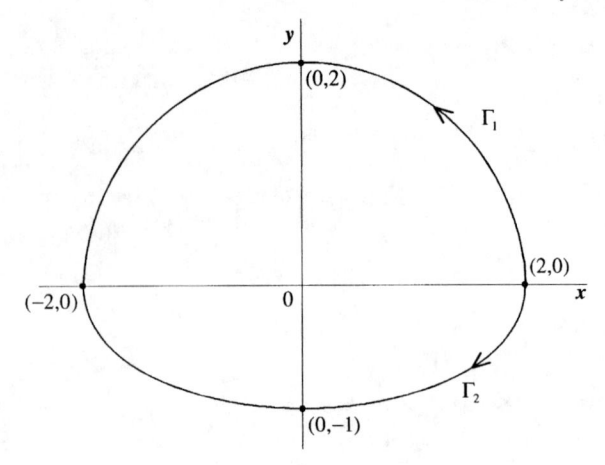

Figure 19.2

have

$$\int_{\Gamma_1} F = \int_0^\pi F(P_1(t)) \cdot P_1'(t)\, dt$$
$$= \int_0^\pi (2 \cdot 2\cos t \cdot 2\sin t, 4\cos^2 t) \cdot (-2\sin t, 2\cos t)\, dt$$
$$= \int_0^\pi (-16 \cos t \sin^2 t + 8 \cos^3 t)\, dt$$
$$= \int (-16 u^2 + 8(1-u^2))\, du, \qquad \begin{aligned} u &= \sin t \\ du &= \cos t\, dt \end{aligned}$$
$$= -\frac{24 u^3}{3} + 8u$$
$$= -8\sin^3 t \big]_0^\pi + 8 \sin t \big]_0^\pi = 0.$$

Also

$$\int_{\Gamma_2} F = \int_0^\pi F(P_2(t)) \cdot P_2'(t)\, dt$$
$$= \int_0^\pi (2 \cdot 2\cos t \cdot (-\sin t), 4\cos^2 t)) \cdot (-2\sin t, -\cos t)\, dt$$
$$= \int_0^\pi (8\cos t \sin^2 t - 4\cos^3 t)\, dt$$
$$= 0.$$

When two line integrals joining the same points have the same value, one suspects that the function being integrated may have a potential. How does

The Fundamental Theorem of Calculus

one know if a function has a potential and how does one find it? Clearly if

$$\nabla \phi(x,y) = \left(\frac{\partial \phi}{\partial x}, \frac{\partial \phi}{\partial y}\right) = (f,g) = F$$

then $\dfrac{\partial \phi}{\partial x} = f$ and $\dfrac{\partial \phi}{\partial y} = g$. If, moreover, f and g both have continuous first order partial derivatives then

$$\frac{\partial^2 \phi}{\partial y \partial x} = \frac{\partial f}{\partial y} \quad \text{and} \quad \frac{\partial^2 \phi}{\partial x \partial y} = \frac{\partial g}{\partial x}.$$

Since $\dfrac{\partial f}{\partial y}$ and $\dfrac{\partial g}{\partial x}$ are both continuous we have

$$\frac{\partial f}{\partial y} = \frac{\partial^2 \phi}{\partial y \partial x} = \frac{\partial^2 \phi}{\partial x \partial y} = \frac{\partial g}{\partial x}.$$

The converse is also true for **appropriate domains** and we will indicate how this can be proved later. Hence, if U is an appropriate open set and $F = (f,g) : U \to \mathbf{R}^2$ has continuous first order partial derivatives then F has a potential if and only if

$$\frac{\partial f}{\partial y} = \frac{\partial g}{\partial x}.$$

In our case $F(x,y) = (2xy, x^2)$. Since

$$\frac{\partial}{\partial y}(2xy) = 2x = \frac{\partial}{\partial x}(x^2)$$

and the set $U = \mathbf{R}^2$ is appropriate it follows that F has a potential ϕ.

This shows that $\int_1 F = \int_2 F$ and could have saved us half our calculations. We would save more calculations, by Theorem 66, if we knew ϕ.

In our example we have $\dfrac{\partial \phi}{\partial x} = 2xy$ and $\dfrac{\partial \phi}{\partial y} = x^2$. If we treat y as a constant, then (1) for functions of one variable tells us that

$$\int \frac{\partial \phi}{\partial x} \, dx = \phi + C$$

where C is some constant of integration. But it is only a constant with respect to x and it may depend on the "constant" y. Hence we write

$$\int \frac{\partial \phi}{\partial x} \, dx = \phi(x,y) + C(y).$$

We evaluate this integral treating y as a constant and obtain

$$\int \frac{\partial \phi}{\partial x} \, dx = \int 2xy \, dx = \frac{2y \cdot x^2}{2} = \phi(x,y) + C(y).$$

Differentiating with respect to y gives
$$x^2 = \frac{\partial \phi}{\partial y} + C'(y) = x^2 + C'(y).$$

Hence $C'(y) = 0$ and C does not depend on y. Since we have seen already that C does not depend on x, it follows that C is a genuine constant. This yields
$$\phi(x,y) = x^2 y + \text{ constant}.$$

We now have a simple answer to our original problem
$$\int_{\Gamma_1} F = \int_{\Gamma_2} F = \phi(-2,0) - \phi(2,0) = 0.$$

The same method can be used to find the harmonic conjugate of a harmonic function.

We briefly mention one useful role (in fact there are many) of line integrals in physics. A **force** or **force field** acting on a subset U of $\mathbf{R}^2$ is a function $F : U \subset \mathbf{R}^2 \to \mathbf{R}^2$. An object placed at the point (x,y) will be subjected to the force $f(x,y)$ in the x-direction and $g(x,y)$ in the y-direction where $F(x,y) = (f(x,y), g(x,y))$. If A and B are points in U and Γ is a curve joining A and B, then the **work** done by the force in moving a particle of unit mass from A to B along the curve Γ is defined as
$$W = \int_{\Gamma} F \cdot T \, ds$$
where T is the unit tangent. If the value of W only depends on A and B, i.e., if F has a potential, the force is said to be **conservative**.

Exercises

19.1 Find the line integral of the function F along the curve Γ
  (a) $F(x,y) = (x^2 - 2xy, 2xy - y^2)$, where Γ is the parabola $y = x^2$ joining the point $(-1,1)$ to the point $(1,1)$,
  (b) $F(x,y) = (x^2 + y^2, x^2 - y^2)$, where Γ is the ellipse $b^2 x^2 + a^2 y^2 = a^2 b^2$.

19.2 Show, by evaluating two different integrals joining the same points, that
$$G(x,y) = (xe^{xy}, ye^{xy})$$
does not have a potential.

19.3 Which of the following have potentials. Justify your answers and find the potentials that exist.

(a) $F(x,y) = (y\cos xy, x\cos xy)$,
(b) $F(x,y) = (2x + y + 2xe^{x^2}, x + 2ye^{y^2})$,
(c) $F(x,y) = (x^2 \sin xy, y^2 \sin xy)$,
(d) $F(x,y) = (2xy + y^3 + 2x\sin(x^2), x^2 + 3xy^2)$.

19.4 A two-dimensional force field $\vec{F}$ has the form $\vec{F}(x,y) = (axy, x^6 y^2)$ where a is a positive constant. The force moves an object from the origin to the line $x = 1$ along a curve of the form $y = bx^c$ where b and c are positive constants. Find b in terms of a if the work done by the force does not depend on c.

19.5 If $u: \mathbf{R}^2 \to \mathbf{R}$ has continuous second order partial derivatives show that $F = (-\dfrac{\partial u}{\partial y}, \dfrac{\partial u}{\partial x})$ has a potential v if and only if u is harmonic. Show that $u + iv$ is $\mathbf{C}$-differentiable. If $u(x,y) = e^x(x\cos y - y\sin y)$, show that u is harmonic, find v and express $u + iv$ as a function of $z = x + iy$.

20

Double Integrals

Summary. We define the double integral of a function over a certain type of domain in $\mathbf{R}^2$. We show, using Fubini's theorem, how to evaluate such integrals and give examples.

We now discuss (**double**) integration of a function of **two** variables $f(x,y)$ over an open set Ω in $\mathbf{R}^2$. Motivated by the one-dimensional theory we partition Ω into rectangles—the natural analogue of intervals—by first drawing horizontal and vertical lines and thus partitioning the x- and y-axes.

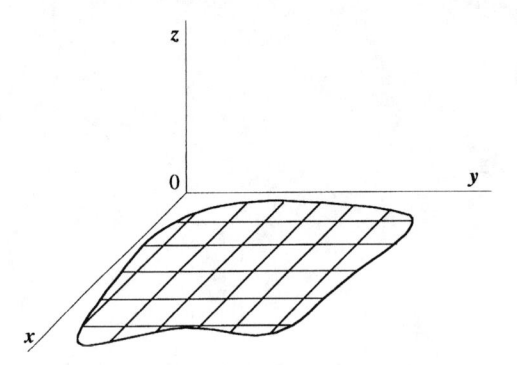

Figure 20.1

We use x_i to denote a typical element of the partition of the x-axis and y_j for the typical element on the y-axis. The resulting grid of rectangles gives us a partition of Ω (Figure 20.1). We again form the Riemann sum

$$\sum_{i,j} f(x_i, y_j) \cdot \Delta x_i \cdot \Delta y_j, \quad \Delta x_i = x_{i+1} - x_i, \quad \Delta y_j = y_{j+1} - y_j$$

where we sum over all rectangles which are strictly contained in Ω. If this

Double Integrals

sum tends to a limit as we take finer and finer partitions we say that f is **integrable** and denote the limit by

$$\iint_\Omega f(x,y)\,dxdy.$$

We call this the **integral** (or **double integral**) of f over Ω. If Ω is the set formed by the inside of a closed curve Γ and f is **continuous** on the set $\Omega \cup \Gamma$ it can be shown that f is integrable over Ω.

Two simple and important examples come to mind immediately. If we let $f(x,y) = 1$ for all x, y then the Riemann sum is just the area of the rectangles in the partition inside Ω and on taking a limit we see that

$$\iint_\Omega dxdy = \textbf{Area of } \Omega.$$

If $f(x,y) \geq 0$ and we try sketch the Riemann sum, we get the situation shown in Figure 20.2 and see that

$$\iint_\Omega f(x,y)\,dx\,dy$$

= **Volume** of the solid over Ω and beneath the graph of f.

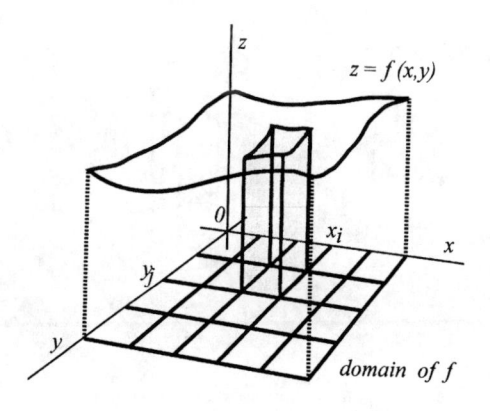

Figure 20.2

We only evaluate such integrals over rather simple domains. A domain is said to be of **type I** if it is bounded above by the graph of a continuous function $y = h(x)$, and bounded below by the graph of a continuous function $y = g(x)$ and on the left and right by vertical lines of finite length (see Figure 20.3).

If we take a fixed interval in the partition of the x-axis, say (x_i, x_{i+1}), and consider the terms in the Riemann sum of f over Ω

$$\sum_{i,j} f(x_i, y_j) \cdot \Delta x_i \cdot \Delta y_j$$

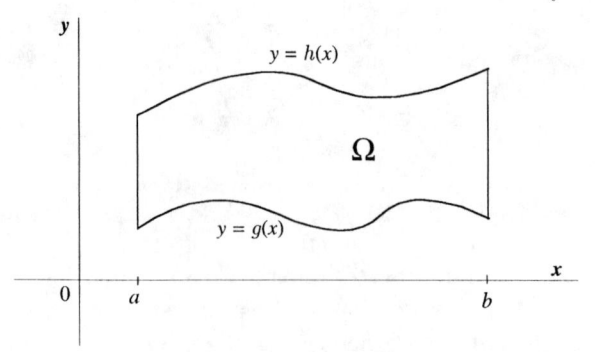

Figure 20.3

which only involve $\Delta x_i = x_{i+1} - x_i$ then we get as in Figure 20.4

$$\left(\sum_j f(x_i, y_j) \cdot \Delta y_j\right) \Delta x_i.$$

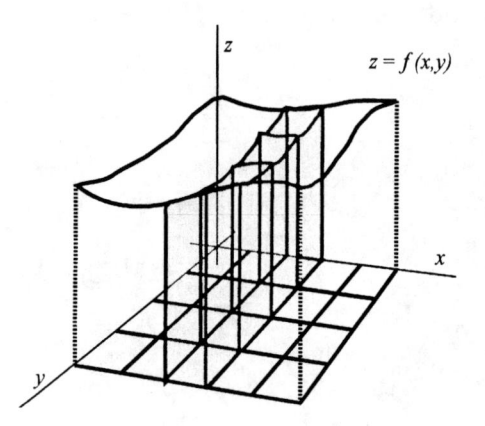

Figure 20.4

Assuming we can first take the limits in the inner sum—this can be justified for **continuous** functions—we get

$$\sum_j f(x_i, y_j) \cdot \Delta y_j \simeq \int_{g(x_i)}^{h(x_i)} f(x_i, y)\, dy.$$

Let

$$H(x) = \int_{g(x)}^{h(x)} f(x, y)\, dy.$$

If we now take the sum over all i then we get the full Riemann sum, i.e.,

Double Integrals

we have

$$\sum_{i,j} f(x_i, y_j) \cdot \Delta x_i \cdot \Delta y_j = \sum_i \left(\sum_j f(x_i, y_j) \cdot \Delta y_j \right) \Delta x_i$$
$$\approx \sum_i H(x_i) \cdot \Delta x_i$$

and taking the limit on both sides we get

$$\int\int_1 f(x,y)\,dydx = \int_a^b H(x)\,dx = \int_a^b \left\{ \int_{g(x)}^{h(x)} f(x,y)\,dy \right\} dx.$$

This is a practical method for evaluating double integrals.

Example 69. We wish to evaluate $\int\int_1 (x^2y + xy^2)\,dxdy$ where Ω is the set of points between the curves $y = x$ and $y = x^2$, $0 \leq x \leq 1$. We have

$$\int\int_1 (x^2y + xy^2)\,dxdy = \int \left\{ \int (x^2y + xy^2)\,dy \right\} dx.$$

To find the limits of integration it is **essential** to sketch the boundaries of Ω (Figure 20.5).

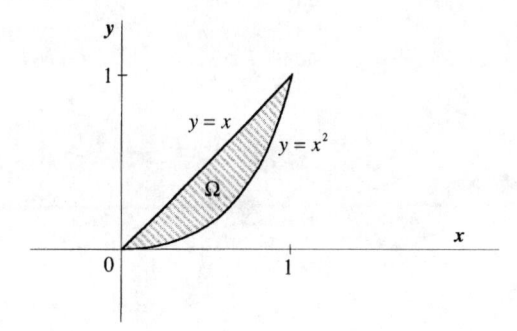

Figure 20.5

We are integrating first over y so x is fixed and y is a variable. The variable y varies along **vertical lines** and is constant on **horizontal lines**. We consider the vertical line through a fixed point x (Figure 20.6). We see that for fixed x, y varies between the values x^2 and x and these give the limits of integration in the inner integral. The inner integral is

$$\int_{x^2}^{x} (x^2y + xy^2)\,dy$$

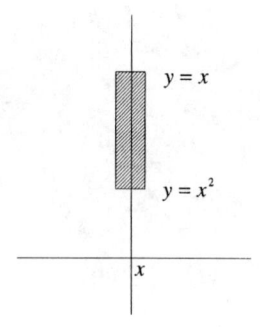

Figure 20.6

and in evaluating this integral we treat x as a constant. The answer should be a function of x **only** since we have integrated out y.

We have

$$\int_{x^2}^{x} (x^2 y + xy^2)\, dy = \left. \frac{x^2 y^2}{2} + \frac{xy^3}{3} \right]_{x^2}^{x}$$

$$= \frac{x^2 x^2}{2} + \frac{xx^3}{3} - \frac{x^2(x^2)^2}{2} - \frac{x(x^2)^3}{3}$$

$$= \frac{5}{6} x^4 - \frac{x^6}{2} - \frac{x^7}{3}.$$

Having integrated out y we see that x varies between 0 and 1 so that the limits of integration of the x variable are 0 and 1. We have

$$\int\!\!\int_1 (x^2 y + xy^2)\, dx dy = \int_0^1 \left\{ \int_{x^2}^{x} (x^2 y + xy^2)\, dy \right\} dx$$

$$= \int_0^1 \left(\frac{5}{6} x^4 - \frac{x^6}{2} - \frac{x^7}{3} \right) dx$$

$$= \left. \frac{x^5}{6} - \frac{x^7}{14} - \frac{x^8}{24} \right]_0^1$$

$$= \frac{1}{6} - \frac{1}{14} - \frac{1}{24} = \frac{3}{56}.$$

This method of integration, together with the similar method got by reversing the roles of x and y, is known as **Fubini's theorem.** We define a domain to be of type II if it is bounded on the left and right by the graphs of continuous functions of y, and above and below by horizontal lines of finite length (Figure 20.7).

The domain Ω is bounded by the graphs of the functions of y, k and l, which are defined on the interval $[c, d]$. If we are given an open set we can recognize that it is of type I if each **vertical** line cuts it at two points except at the end points and type II if each **horizontal** line cuts it at two

Double Integrals   161

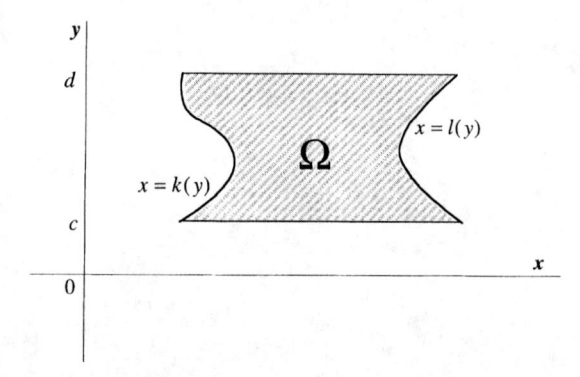

Figure 20.7

points except at the end points. Many domains, for instance the domain in Example 69, are both of type I and type II. When a domain is both of type I and type II one should examine both options. For domains of type II we have

$$\iint_I F(x,y)\,dx\,dy = \int_c^d \left\{ \int_{k(y)}^{l(y)} F(x,y)\,dx \right\} dy.$$

One very important and simple situation occurs when Ω is the rectangle $[a,b] \times [c,d]$ and f is a product of a function of x and a function of y, i.e., $f(x,y) = g(x) \cdot h(y)$. In this case Ω is both of type I and type II and the value of the inner integral is independent of the variables (why?). This means that the double integral is a product of two integrals and we have

$$\iint_I f(x,y)\,dx\,dy = \left(\int_a^b g(x)\,dx \right) \cdot \left(\int_c^d h(y)\,dy \right).$$

These **two conditions**—one on the domain of integration and the other on the function—**must be present**. In some cases we may have $f(x,y) = g(x)$, i.e., f is independent of y. In such a situation take $h(y) = 1$ all y and apply the above (and of course the same remark applies when $f(x,y) = h(y)$).

Example 70. Evaluate

$$\iint_I \frac{x}{\sqrt{16+y^7}}\,dx\,dy$$

over the set Ω bounded above by the line $y = 2$, below by the graph of $y = x^{1/3}$ and on the left by the y-axis (Figure 20.8). By inspection the domain Ω is of type I and type II and we have a choice of method, i.e., we can integrate first with respect to either variable. Our choice may be important since one method may be very simple and the other quite difficult.

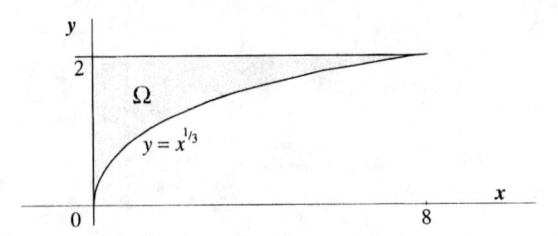

Figure 20.8

We have to evaluate two integrals of a single variable. In the first integral, the inner integral, one of the variables takes a fixed value and is really a constant. Thus we have to first evaluate either

$$\int x \, dx \quad \text{or} \quad \int \frac{dy}{\sqrt{16 + y^7}}.$$

In these situations looks are usually not deceiving and we opt for the simpler looking integral. So, we choose to integrate first with respect to x. The limits of integration in the first integral will influence the degree of difficulty that arises in evaluating the second, or outer, integral. If you run into problems with the second integral you should consider starting again using the other variable. In our case we have decided to consider

$$\int \left\{ \int \frac{x}{\sqrt{16 + y^7}} \, dx \right\} dy$$

and now we need to determine the variation in x for fixed y. We draw a typical line through Ω on which y is constant—i.e., a horizontal line.

We need to express the end points in terms of y. Using once more Figure 20.8 we see that the end points of the line of variation of x are $(0, y)$ and (x, y) where $y = x^{1/3}$. Hence $y^3 = x$ and we have the required variation of x (Figure 20.9).

Variation of x

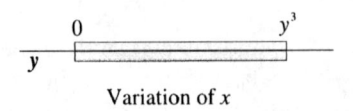

Figure 20.9

Double Integrals

We see also that y varies from 0 to 2. Hence

$$\iint_\Omega \frac{x}{\sqrt{16+y^7}}\,dx\,dy = \int_0^2 \left\{ \int_0^{y^3} \frac{x\,dx}{\sqrt{16+y^7}} \right\} dy$$

$$= \int_0^2 \frac{x^2\,dx}{2\sqrt{16+y^7}} \bigg]_0^{y^3} dy$$

$$= \int_0^2 \frac{y^6}{2\sqrt{16+y^7}}\,dy$$

$$= \frac{1}{14}\int \frac{dw}{\sqrt{w}} \qquad \begin{aligned} w &= 16+y^7 \\ dw &= 7y^6\,dy \end{aligned}$$

$$= \frac{1}{14} \cdot \frac{w^{1/2}}{1/2} = \frac{1}{7}(16+y^7)^{1/2}\bigg]_0^2$$

$$= \frac{1}{7}((144)^{1/2} - (16)^{1/2}) = \frac{8}{7}.$$

Example 71. In this example we wish to reverse the order of integration and evaluate

$$\int_0^3 \left\{ \int_1^{\sqrt{4-y}} (x+y)\,dx \right\} dy.$$

The limits of integration in the inner integral show that the left-hand side of the domain of integration, Ω, is bounded by the line $x = 1$ and the right-hand side by points satisfying $x = \sqrt{4-y}$.

We have x in terms of y because the right-hand side of our diagram is the graph of a function of y. We are more familiar with the graph of a function f of x and denote a typical point by (x,y) or $(x, f(x))$. If it is also the graph of a function of y then x can be expressed as a function of y, i.e., $x = g(y)$. The function g is the inverse of f and f is the inverse of g. We find g by using the equation $y = f(x)$ to solve for x in terms of y and conversely, as is the case here, we use $x = g(y)$ to find f. We have $x = \sqrt{4-y}$, i.e., $x^2 = 4-y$ and $y = 4-x^2$. A typical point on such a graph can be written in three ways; $(x, 4-x^2) = (x,y) = (\sqrt{4-y}, y)$ (see Figure 20.10). This method of labelling graphs is important here and elsewhere.

We have shown that the top of Ω is bounded by the graph of $y = 4-x^2$. Reversing the order of integration we obtain

$$\int_1^2 \left\{ \int_0^{4-x^2} (x+y)\,dy \right\} dx = \int_1^2 \left(xy + \frac{y^2}{2} \right)\bigg]_0^{4-x^2} dx$$

$$= \int_1^2 \left(x(4-x^2) + \frac{(4-x^2)^2}{2} \right) dx = \frac{241}{60} \quad \text{(eventually)}.$$

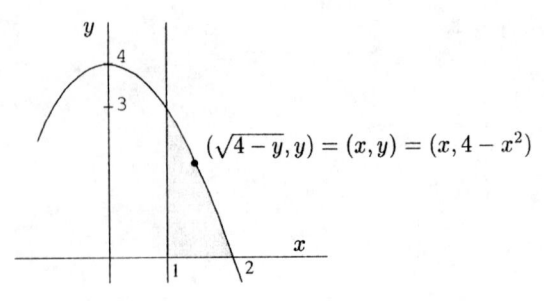

Figure 20.10

$(\sqrt{4-y}, y) = (x, y) = (x, 4-x^2)$

Exercises

20.1 Evaluate $\displaystyle\iint_A x^2 \sin^2 y \, dx dy$ where
$$A = \{(x, y) \in \mathbf{R}^2; 0 \le x \le 1, 0 \le y \le \frac{\pi}{4}\}.$$

20.2 Find $\displaystyle\iint_A x \cos(x+y) \, dx dy$ where A is the area bounded by the triangle with vertices $(0, 0)$, $(\pi, 0)$ and (π, π).

20.3 Let R be the region in $\mathbf{R}^2$ bounded by the graphs of $y = x^2$ and $y = 2x$. Evaluate
$$\iint_R (x^3 + 4y) \, dx dy.$$

20.4 Reverse the order of integration and evaluate
$$\int_0^1 \left\{ \int_{2x}^2 e^{y^2} \, dy \right\} dx$$
$$\int_1^e \left\{ \int_0^{\log y} x \, dx \right\} dy.$$

20.5 If $\lambda > 0$ and $f(x, y) = \lambda^2 e^{-\lambda y}$ for $0 < x < y$ (and is otherwise zero) show that
$$\iint_{\mathbf{R}^2} e^{tx + sy} f(x, y) \, dx \, dy = \frac{\lambda^2}{(\lambda - t - s)(\lambda - s)}$$
for $s < \lambda$ and $t + s < \lambda$. This function of s and t is called the **moment generating function** of f.

21

Coordinate Systems

Summary. *We discuss coordinate systems. We show that different parametrizations lead to the same value of a line integral along a directed curve. We prove the change of variable rule for double integrals. We discuss Cartesian and polar coordinates and show by example how to translate from one system to the other.*

A method of identifying each element of a set by a number or by a pair of numbers or even by a set of numbers is called a **coordinate system**. The plane $\mathbf{R}^2$, which consists of all pairs of real numbers (x, y), comes to us with a ready made coordinate system, the **Cartesian coordinate system**. Sets admit, however, many different coordinate systems. A point in $\mathbf{R}^2$ can also be identified by its distance from the origin, r, and the angle, θ, that the line joining it to the origin makes with the positive x-axis (Figure 21.1).

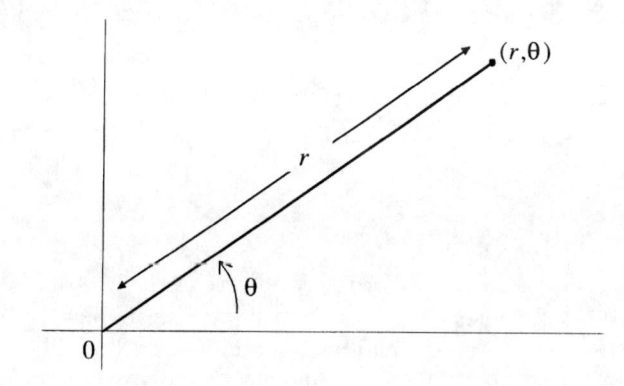

Figure 21.1

We call (r, θ) the **polar coordinates** of the point.

Since points are completely identified by their coordinates in any coordinate system, it is possible to go from one set of coordinates to another. For example, if (r, θ) are the polar coordinates of a point P in $\mathbf{R}^2$ then $x = r\cos\theta$ and $y = r\sin\theta$ give the Cartesian coordinates (x, y) of the point P.

If $F = (f, g) : \mathbf{R}^2 \to \mathbf{R}^2$ is an injective (i.e., one to one) mapping then the pair $\big(f(x,y), g(x,y)\big)$ gives the coordinates of a new coordinate system on $\mathbf{R}^2$. This is one of the standard methods of generating new coordinate systems from old ones. A function on a set may be described in terms of **any** coordinate system and **in different systems** the function may **appear** very different. For example, if we consider the function that measures the distance of a point in $\mathbf{R}^2$ from the origin then in Cartesian coordinates we have the description

$$f(x, y) = (x^2 + y^2)^{1/2}$$

while in polar coordinates we get

$$g(r, \theta) = r.$$

The functions f and g appear very different but in fact they are the **same** function expressed in **different** coordinate systems.

To go from one system to the other we use $x = r\cos\theta$ and $y = r\sin\theta$ and get

$$(x^2 + y^2)^{1/2} = (r^2\cos^2\theta + r^2\sin^2\theta)^{1/2} = r\ .$$

For this reason changing coordinates is often called a **change of variable**. Our approach to a particular problem will often depend on the form in which it is presented. For example, the forms of f and g above may suggest different approaches. We use different coordinate systems in order to find one in which we can solve, or more easily solve, a given problem.

The mappings that pick out a particular coordinate in a coordinate system are called the **coordinate mappings**. For Cartesian coordinates these are

$$(x, y) \to x \quad \text{and} \quad (x, y) \to y.$$

We already used these in Chapter 6 when discussing continuity. For polar coordinates these are the mappings

$$(r, \theta) \to r \quad \text{and} \quad (r, \theta) \to \theta.$$

Information on the structure of the level sets of the coordinate mappings can also be helpful when choosing a coordinate system. For Cartesian coordinates these are the vertical and horizontal lines, and for polar coordinates we get circles centred at the origin and rays originating at the origin (Figure 21.2).

Coordinate Systems

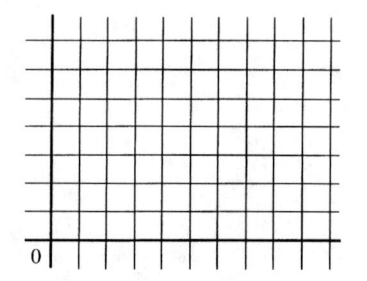

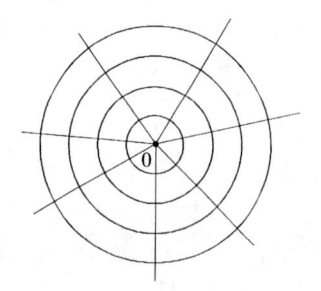

Figure 21.2

It is not surprising that polar coordinates are useful with problems involving circular regions and with functions that involve the expression $x^2 + y^2$, when described in Cartesian coordinates.

To be able to use different coordinate systems we must be able to **translate** from one system to another. We begin our study in this direction by looking at curves and line integrals. If $P : [a, b] \to \Gamma$ is a parametrization of a directed curve Γ, then each point t in the interval $[a, b]$ determines a unique point $P(t)$ on the curve Γ and thus we may consider $[a, b]$ as a coordinate system for Γ. If $F = (f, g) : \mathbf{R}^2 \to \mathbf{R}^2$ is defined on an open set containing Γ, then we defined

$$\int_\Gamma F \quad \text{as} \quad \int_a^b F(P(t)) \cdot P'(t)\, dt.$$

We now wish to show that two different parametrizations (i.e., two different coordinate systems) on Γ give us the same value of the integral.

So suppose $P_1 : [c, d] \to \Gamma$ is another parametrization of Γ. We wish to show that

$$\int_a^b F(P(t)) \cdot P'(t)\, dt = \int_c^d F(P_1(t)) \cdot P_1'(t)\, dt.$$

To prove this we recall the one variable rule for change of variable in integrals.

If $g : [c, d] \to [a, b]$ is a strictly **monotonic**, i.e., either strictly increasing or strictly decreasing, differentiable mapping from $[c, d]$ onto $[a, b]$ and f is a continuous function on $[a, b]$ then

$$\int_a^b f(x)\, dx = \int_c^d f(g(y)) |g'(y)|\, dy.$$

Let $P(t) = (x(t), y(t))$ and let s denote the length function associated with

P. If $F = (f, g)$ then

$$\int_a^b F(P(t)) \cdot P'(t) \, dt$$

$$= \int_a^b \left(f(P(t)) \cdot x'(t) + g(P(t)) \cdot y'(t) \right) dt$$

$$= \int_0^l \left(f(P \circ s^{-1}(w)) x'(s^{-1}(w)) + g(P \circ s^{-1}(w)) y'(s^{-1}(w)) \right) (s^{-1})'(w) \, dw$$

(by the change of variable $t = s^{-1}(w)$)

$$= \int_0^l \left(f(P \circ s^{-1}(w)) \cdot (x \circ s^{-1})'(w) + g(P \circ s^{-1}(w)) \cdot (y \circ s^{-1})'(w) \right) dw$$

$$= \int_0^l \left(F(P \circ s^{-1}(w)) \cdot (P \circ s^{-1})'(w) \right) dw.$$

Let s_1 denote the length function associated with P_1. Since $P \circ s^{-1}$ and $P_1 \circ s_1^{-1}$ are both unit speed parametrizations of Γ on $[0, l]$, where l is the length of Γ, it follows that $P \circ s^{-1} = P_1 \circ s_1^{-1}$. Hence

$$\int_a^b F(P(t)) \cdot P'(t) \, dw = \int_0^l F((P \circ s^{-1})(w)) \cdot (P \circ s^{-1})'(w) \, dw$$

$$= \int_0^l F((P_1 \circ s_1^{-1})(w)) \cdot (P_1 \circ s_1^{-1})'(w) \, dw$$

$$= \int_c^d F(P_1(t)) \cdot P_1'(t) \, dt.$$

This shows that the line integral does not depend on the parametrization.

Example 72. We continue translating problems from one coordinate system to another. We have defined a function f to be **harmonic** if

$$\frac{\partial^2 f}{\partial x^2} + \frac{\partial^2 f}{\partial y^2} = 0$$

where (x, y) are the Cartesian coordinates. If we are given a function in polar coordinates we can check whether or not it is harmonic by changing the coordinate system and using the above. We would prefer to be able to check this property directly using polar coordinates. We will use the chain rule to find derivatives. Consider the composition

$$(r, \theta) \quad \rightarrow \quad (r \cos \theta, r \sin \theta)$$
$$\parallel$$
$$(x, y) \quad \rightarrow \quad f(x, y)$$

Coordinate Systems

and let $g(r, \theta) = f(r \cos \theta, r \sin \theta)$. By the chain rule

$$\frac{\partial g}{\partial r} = \frac{\partial f}{\partial x} \cos \theta + \frac{\partial f}{\partial y} \sin \theta$$

$$\frac{\partial^2 g}{\partial r^2} = \frac{\partial^2 f}{\partial x^2} \cos^2 \theta + 2 \frac{\partial^2 f}{\partial x \partial y} \cos \theta \sin \theta + \frac{\partial^2 f}{\partial y^2} \sin^2 \theta$$

$$\frac{\partial g}{\partial \theta} = \frac{\partial f}{\partial x}(-r \sin \theta) + \frac{\partial f}{\partial y} r \cos \theta$$

$$\frac{\partial^2 g}{\partial \theta^2} = \frac{\partial^2 f}{\partial x^2} r^2 \sin^2 \theta - 2 \frac{\partial^2 f}{\partial x \partial y} r \sin \theta \cos \theta + \frac{\partial^2 f}{\partial y^2} r^2 \cos^2 \theta$$
$$+ \frac{\partial f}{\partial x}(-r \cos \theta) - \frac{\partial f}{\partial y} r \sin \theta.$$

Hence

$$\frac{\partial^2 g}{\partial r^2} + \frac{1}{r^2} \frac{\partial^2 g}{\partial \theta^2} + \frac{1}{r} \frac{\partial g}{\partial r} = \frac{\partial^2 f}{\partial x^2}(\cos^2 \theta + \sin^2 \theta) + \frac{\partial^2 f}{\partial y^2}(\cos^2 \theta + \sin^2 \theta)$$
$$= \frac{\partial^2 f}{\partial x^2} + \frac{\partial^2 f}{\partial y^2}$$

and $g(r, \theta)$ is harmonic if and only if it satisfies the equation

$$\frac{\partial^2 g}{\partial r^2} + \frac{1}{r^2} \frac{\partial^2 g}{\partial \theta^2} + \frac{1}{r} \frac{\partial g}{\partial r} = 0.$$

For example, if $f(z) = z^3$, then

$$(re^{i\theta})^3 = r^3 e^{i3\theta} = r^3 \cos 3\theta + ir^3 \sin 3\theta.$$

Hence $u(r, \theta) = r^3 \cos 3\theta$ is the real part of f in polar coordinates. Since

$$\frac{\partial^2 u}{\partial r^2} + \frac{1}{r^2} \frac{\partial^2 u}{\partial \theta^2} + \frac{1}{r} \frac{\partial u}{\partial r} = 6r \cos 3\theta - \frac{1}{r^2} \cdot 9r^3 \cos \theta + \frac{1}{r} \cdot 3r^2 \cos 3\theta$$
$$= (6r - 9r + 3r) \cos 3\theta$$
$$= 0$$

we see that u is a harmonic function. We have already proved this result in Example 61 using Cartesian coordinates. The Cauchy-Riemann equations in polar form are given in Exercise 21.5.

We now discuss changing coordinates in double integrals and obtain a rule similar to the one variable rule quoted earlier. In motivating double integrals we partitioned the domain of integration by using vertical and horizontal lines. Any partition will do as long as the subdivisions tend to zero as we take further refinements. We consider an integrable function f on Ω and a given coordinate system (u, v). If we consider another coordinate system (s, t) and a mapping g which transforms U in $\mathbf{R}^2_{(s,t)}$ onto Ω in $\mathbf{R}^2_{(u,v)}$

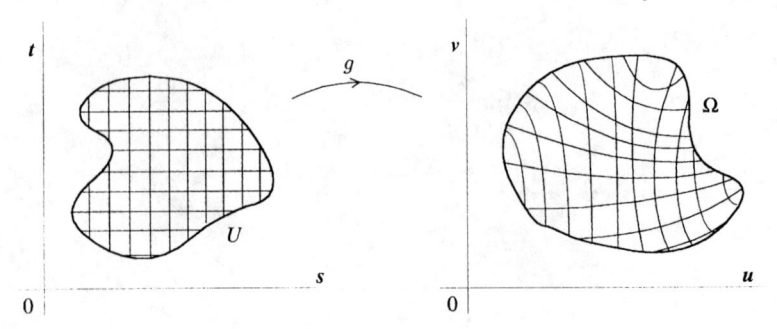

Figure 21.3

in a smooth bijective fashion, then any partition of U will give rise to a partition of Ω (see Figure 21.3).

To approximate
$$\int\int_V f(u,v)\, du\, dv$$
we first form a rectangular partition of U, transfer it to $\Omega = g(U)$, where it is still a partition, not necessarily composed of rectangles, and then form a Riemann sum for f
$$\sum_{i,j} f(g(s_i, t_j))\mathrm{Area}(g(\Delta s_i \times \Delta t_j)).$$

A typical term in the Riemann sum is obtained from Figure 21.4 where $g(A) = A'$, $g(B) = B'$, $g(C) = C'$ and $g(D) = D'$.

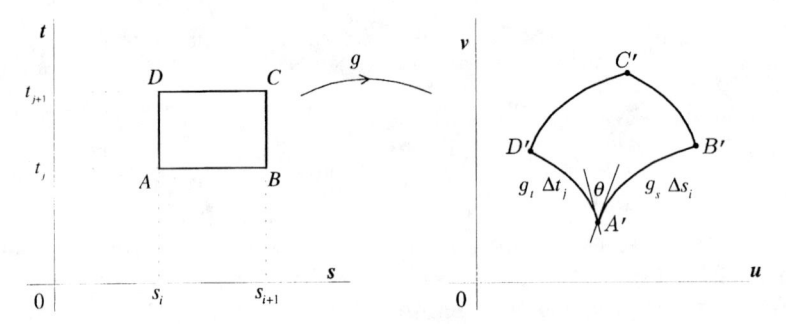

Figure 21.4

We have
$$\int\int_V f(u,v)\, du\, dv \approx \sum_{i,j} f(g(s_i,t_j)) \cdot \mathrm{Area}(A'B'C'D').$$

We recall (see Figure 21.5) how to calculate the area of the parallelogram generated by two vectors in $\mathbf{R}^2$, (a_1, b_1) and (a_2, b_2).

Coordinate Systems

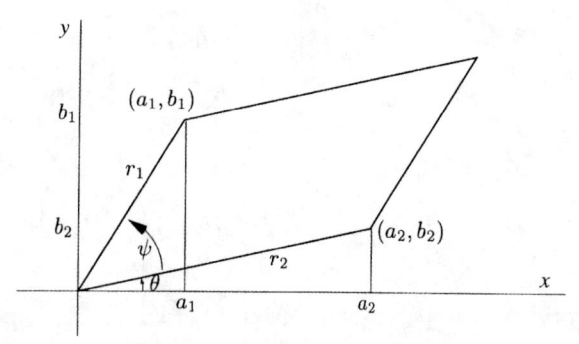

Figure 21.5

$$
\begin{aligned}
\text{Area} &= |r_1 r_2 \sin \psi| \\
&= |r_1 r_2 \sin(\theta + \psi - \theta)| \\
&= |r_1 r_2 (\sin(\theta + \psi)\cos\theta - \cos(\theta + \psi)\sin\theta)| \\
&= \left|r_1 r_2 \left(\frac{b_1}{r_1} \cdot \frac{a_2}{r_2} - \frac{a_1}{r_1} \cdot \frac{b_2}{r_2}\right)\right| \\
&= |b_1 a_2 - a_1 b_2| \\
&= \left|\det\begin{pmatrix} a_1 & a_2 \\ b_1 & b_2 \end{pmatrix}\right|.
\end{aligned}
$$

Since $A'B'$ is the curve parametrized by

$$s \in [s_i, s_{i+1}] \to g(s, t_j)$$

we have $A'B' \approx g_s \Delta s_i$. This approximation was used in evaluating the length of a curve in Chapter 13. Similarly $A'D' \approx g_t \Delta t_j$. In our case $(a_1, b_1) = g_s \Delta s_i$ and $(a_2, b_2) = g_t \Delta t_j$. Hence,

$$
\begin{aligned}
\text{Area}(A'B'C'D') &= |\det(g_s, g_t)| \Delta s_i \Delta t_j \\
&= |\det(g')| \Delta s_i \Delta t_j.
\end{aligned}
$$

The notation $\dfrac{\partial(u,v)}{\partial(s,t)}$ and $J(g)$ are also used in place of $\det(g')$ and this determinant is called the **Jacobian** of g.

Returning to the approximation we see that

$$\iint f(u,v)\, du\, dv \approx \sum_{i,j} f\big(g(s_i, t_j)\big)\Big|\det\big(g'(s_i, t_j)\big)\Big|\Delta s_i \Delta t_j$$

and taking the limit of the right-hand side we get the following proposition. Our analysis did not fully uncover the **regularity conditions** required to change variables. We include them, without proof, in the statement of Proposition 73.

Proposition 73. *(Change of variables formula for double integrals).* If U is an open subset of $\mathbf{R}^2$, $g : U \to \Omega \subset \mathbf{R}^2$ is a bijective (i.e., one-to-one and onto) function with continuous first order partial derivatives such that $\det(g') \neq 0$ and f is a continous integrable function on Ω, then Ω is open and

$$\int\int f(u,v)\, du\, dv = \int\int_U f(g(s,t)) \left|\det(g'(s,t))\right| ds\, dt.$$

Example 74. We use polar coordinates to evaluate

$$\int\int_{\substack{x^2+y^2<1 \\ y>0}} \cos\left((x^2+y^2)^{1/2}\right) dx\, dy.$$

We change coordinates (Figure 21.6) by using the mapping $(r,\theta) \to (x,y)$. Since $g(r,\theta) = (r\cos\theta, r\sin\theta)$,

$$g'(r,\theta) = \begin{pmatrix} \cos\theta & -r\sin\theta \\ \sin\theta & r\cos\theta \end{pmatrix},$$

and $\det(g'(r,\theta)) = r\cos^2\theta + r\sin^2\theta = r$. In our case

$$U = \{(r,\theta); 0 \le r \le 1, 0 \le \theta \le \pi\}$$

and

$$\Omega = g(U) = \{(x,y); x^2 + y^2 < 1, y > 0\}$$
$$= \text{upper half of the unit disc.}$$

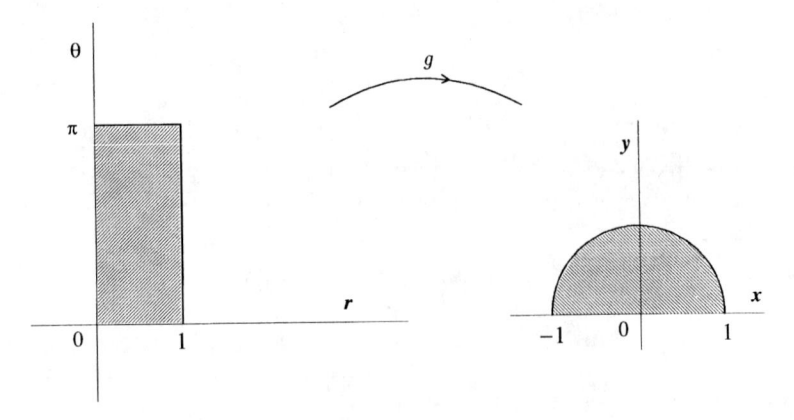

Figure 21.6

Hence

$$\iint_{\substack{x^2+y^2\le 1\\ y>0}} \cos((x^2+y^2)^{1/2})\,dx\,dy = \iint_{[0,1]\times[0,\pi]} r\cos r\,dr\,d\theta$$

$$= \int_0^1 \left\{\int_0^\pi r\cos r\,d\theta\right\}dr$$

$$= \pi \int_0^1 r\cos r\,dr$$

$$= \pi \int_0^1 r\frac{d}{dr}(\sin r)\,dr$$

$$= \pi r\sin r\Big]_0^1 - \pi \int_0^1 \sin t\,dt$$

$$= \pi(\sin 1 + \cos 1) - \pi.$$

Exercises

21.1 By changing to polar coordinates evaluate the following:

(a) $$\int_0^1 \left\{\int_0^{\sqrt{1-x^2}} \exp(\sqrt{x^2+y^2})\,dy\right\}dx$$

(b) $$\iint_A \frac{x^2}{x^2+y^2}\,dx\,dy$$

where

$$A = \{(x,y) \in \mathbf{R}^2; a^2 \le x^2+y^2 \le b^2\}$$

$a, b \in \mathbf{R}$ and $0 < a < b$

(c) $$\iint_A (x^2+y^2)\,dx\,dy$$

where $A = \{(x,y) \in \mathbf{R}; x^2+y^2 \le 2y\}$.

21.2 If f and g are two differentiable mappings show that

$$J(f\circ g) = J(f)J(g).$$

If $(x,y) \to (u,v)$ is a change of variable show that

$$\left(\frac{\partial(u,v)}{\partial(x,y)}\right)^{-1} = \frac{\partial(x,y)}{\partial(u,v)}.$$

21.3 Let R denote the square $ABCD$ where $A = (0,0)$, $B = (1,1)$, $C = (2,0)$

and $D = (1, -1)$. By using the change of variable $(u, v) \to (x, y)$, where $x = \frac{1}{2}(u+v)$ and $y = \frac{1}{2}(u-v)$, evaluate

$$\int\int_R x^2 y^2 \, dx \, dy.$$

21.4 If $f = u + iv$ is a holomorphic function show that uv and $e^u \cos v$ are harmonic functions.

21.5 If the real and imaginary parts, u and v, respectively, of a function $f : \mathbf{C} \to \mathbf{C}$, with continuous first order partial derivatives, are given in polar form show that f is $\mathbf{C}$-differentiable (or holomorphic) if and only if u and v satisfy

$$\frac{\partial u}{\partial r} = \frac{1}{r}\frac{\partial v}{\partial \theta} \quad \text{and} \quad \frac{\partial v}{\partial r} = -\frac{1}{r}\frac{\partial u}{\partial \theta}.$$

(These are the Cauchy-Riemann equations in polar form.) Find the real and imaginary parts of $f(z) = ze^z$ in polar form and show that they satisfy the above equations.

21.6 Let $f : U \subset \mathbf{C} \to \mathbf{C}$, $f = u + iv$, denote an injective holomorphic mapping. Let $g : U \subset \mathbf{R}^2 \to \mathbf{R}^2$ be the change of coordinates given by $(x, y) \to (u, v)$. Use the Cauchy-Riemann equations to show that $J(g) = |f'|^2$.

Let $f(z) = \frac{1}{2}(z + z^{-1})$. Show that f maps

$$\{(x, y); 1 < x^2 + y^2 < 2, x > 0, y > 0\}$$

onto the set

$$\{(u, v); 144u^2 + 400v^2 < 225, u > 0, v > 0\}.$$

Sketch both of these sets (see also Exercise 17.6).

22
Green's Theorem

Summary. *We prove Green's theorem and thus obtain a two-dimensional version of the fundamental theorem of calculus. Applications of Green's theorem to the evaluation of integrals, to complex analysis (Cauchy's theorem), to the existence of a potential and to vector calculus (the divergence theorem) are given.*

In this chapter we prove a double integral version of
$$\int_a^b \frac{d}{dx} f(x)\, dx = f(b) - f(a)$$
known as Green's theorem. This theorem together with its generalization to higher dimensions, known as **Stokes' theorem**, is one of the most useful theorems in mathematics. In the two-dimensional case we find a wide variety of applications.

Green's theorem. *If U is an open subset of $\mathbf{R}^2$ and Γ is a closed curve such that Γ and Ω (the interior of Γ) both lie in U, then*
$$\oint_\Gamma M\, dx + N\, dy = \int\int_\Omega \left(\frac{\partial N}{\partial x} - \frac{\partial M}{\partial y} \right) dx dy$$
for any functions M and N with continuous first order partial derivatives on U.

We have used the notation $\oint$ to remind us that we are integrating around a closed curve. By **definition** this is directed **anticlockwise**. To make the above rigorous, it is necessary to give mathematical precision to the terms 'inside of a curve' and 'anticlockwise'. At first glance this appears simple but in fact it is very far from being the case. In dealing with concrete examples there is usually no difficulty but the general definition is much more subtle. Many advances in mathematics can be traced to the time when these and similar everyday terms were given precise mathematical definitions.

Proof of Green's theorem (for a domain Ω, which is both of type I and type II).

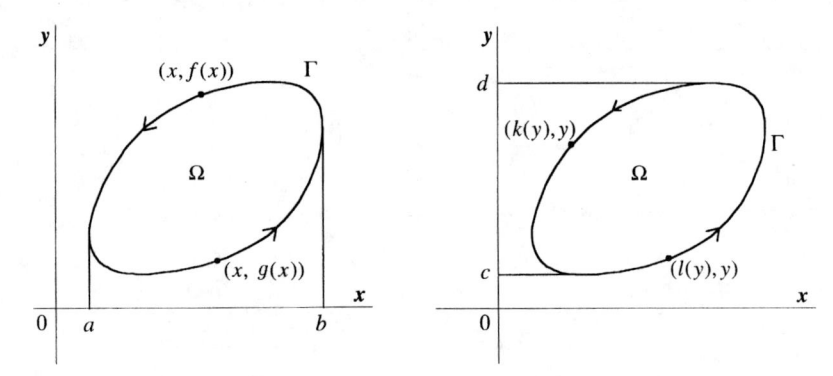

Figure 22.1

Using the type II representation we have

$$\iint \frac{\partial N}{\partial x} dx dy = \int_c^d \left\{ \int_{k(y)}^{l(y)} \frac{\partial N}{\partial x} dx \right\} dy.$$

By the fundamental theorem of calculus for one variable we have

$$\int_{k(y)}^{l(y)} \frac{\partial N}{\partial x}(x,y) \, dx = N(x,y) \Big]_{k(y)}^{l(y)} = N(l(y),y) - N(k(y),y)$$

and so

$$\iint \frac{\partial N}{\partial x} dx dy = \int_c^d \left(N(l(y),y) - N(k(y),y) \right) dy.$$

On the other hand, using the direction on the graphs as indicated in Figure 22.1, we have

$$\oint_\Gamma N \, dy = \int_{\text{(Graph of } l\text{)}} N \, dy - \int_{\text{(Graph of } k\text{)}} N \, dy$$

$$= \int_c^d N(l(y),y) \, dy - \int_c^d N(k(y),y) \, dy$$

and we have shown

$$\iint \frac{\partial N}{\partial x} dx dy = \oint_\Gamma N \, dy.$$

Similarly

$$\iint \frac{\partial M}{\partial y} dx dy = \int_a^b \left\{ \int_{g(x)}^{f(x)} \frac{\partial M}{\partial y} dy \right\} dx$$
$$= \int_a^b \Big(M(x, f(x)) - M(x, g(x)) \Big) dx$$

and

$$\oint_1 M\, dx = \int_{\text{(Graph of } g)} M\, dx - \int_{\text{(Graph of } f)} M\, dx$$
$$= \int_a^b M(x, g(x))\, dx - \int_a^b M(x, f(x))\, dx$$
$$= -\iint \frac{\partial M}{\partial y} dx dy.$$

This proves the theorem.

Example 75. If $M(x,y) = y$ then $\frac{\partial M}{\partial y} = 1$ and if $N(x,y) = x$ then $\frac{\partial N}{\partial x} = 1$. Hence

$$\iint dx\,dy = -\oint_1 y\,dx \quad (\text{Let } M(x,y) = y \text{ and } N(x,y) = 0)$$
$$= \oint_1 x\,dy \quad (\text{Let } M(x,y) = 0 \text{ and } N(x,y) = x)$$
$$= \text{Area of } \Omega.$$

We can combine these to get

$$\text{Area } (\Omega) = \frac{1}{2} \oint_1 x\,dy - y\,dx.$$

Let Ω denote the interior of the ellipse $\frac{x^2}{a^2} + \frac{y^2}{b^2} = 1$, i.e.,

$$\Omega = \{(x,y); \frac{x^2}{a^2} + \frac{y^2}{b^2} < 1\}.$$

The mapping $P(t) = (a\cos t, b\sin t)$, $0 \le t \le 2\pi$, is an anticlockwise parametrization of the boundary curve Γ. We have

$$\text{Area}(\Omega) = \oint_\Gamma x\,dy$$
$$= \int_0^{2\pi} a\cos(t) \cdot b\cos(t)\,dt$$
$$= ab\int_0^{2\pi} \cos^2 t\,dt = \pi ab$$

Example 76. Let Γ denote the closed curve bounded by $y^2 = x$ and $y = -x$ between $(0,0)$ and $(1,-1)$. We wish to use Green's theorem to evaluate $\oint_\Gamma (x^2+y)\,dx + xy^2\,dy$. We begin by sketching the curves involved (Figure 22.2).

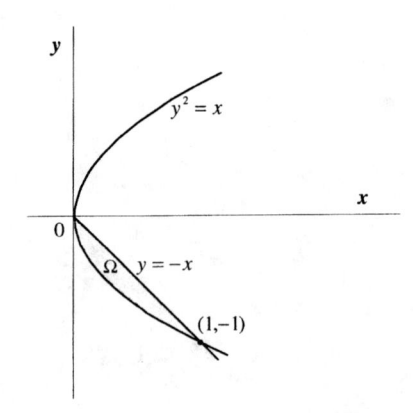

Figure 22.2

By Green's theorem, taking $M(x,y) = x^2 + y$ and $N(x,y) = xy^2$ we have

$$\oint_\Gamma (x^2+y)dx + xy^2\,dy = \int\int (y^2 - 1)\,dx dy.$$

We can evaluate this by working out either of the following:

$$\int\left\{\int (y^2-1)dy\right\}dx \quad \text{or} \quad \int\left\{\int (y^2-1)dx\right\}dy.$$

The second appears simpler. As illustrated in Figure 22.3, in the inner

Green's Theorem

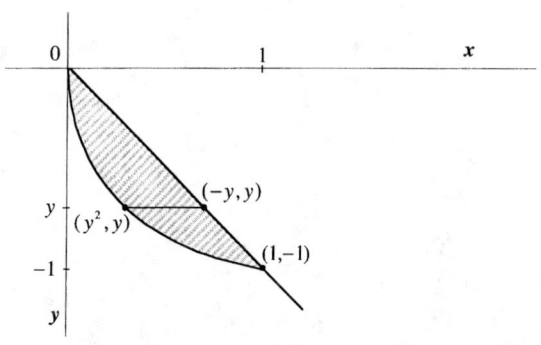

Figure 22.3

integral x is the variable and

$$\int_{-1}^{0}\left\{\int_{y^2}^{-y}(y^2-1)dx\right\}dy = \int_{-1}^{0}(y^2-1)x\Big]_{y^2}^{-y}dy$$

$$= \int_{-1}^{0}(y^2-1)(-y-y^2)dy$$

$$= -\int_{-1}^{0}(y^4+y^3-y^2-y)\,dx$$

$$= -\left(\frac{y^5}{5}+\frac{y^4}{4}-\frac{y^3}{3}-\frac{y^2}{2}\right)\Big]_{-1}^{0}$$

$$= -\frac{1}{5}+\frac{1}{4}+\frac{1}{3}-\frac{1}{2}$$

$$= -\frac{7}{60}.$$

We now turn to the problem of the existence of a potential for the function $F = (f,g)$. We have noted, but not proved, that on appropriate domains we get a positive answer if $\dfrac{\partial f}{\partial y} = \dfrac{\partial g}{\partial x}$. To prove this it suffices, by Theorem 67, to show that line integrals of F along curves in the domain only depend on the end points. We give a partial proof based on our version of Green's theorem, which can be developed into a full proof. We take $U = \mathbf{R}^2$ and fix two arbitrary points A and B. We let Γ_1 and Γ_2 denote two different curves joining A to B such that the set U between them is an open set of type I and type II. Let $\widetilde{\Gamma}_2$ denote the curve obtained from Γ_2 by reversing the direction. Then $\Gamma_1 \cup \widetilde{\Gamma}_2$ is a closed curve (see Figure 22.4). By Green's theorem

$$\oint_{\Gamma_1\cup\widetilde{\Gamma}_2} f\,dx + g\,dy = \int\int_U \left(\frac{\partial g}{\partial x}-\frac{\partial f}{\partial y}\right)dx\,dy = 0.$$

Hence

$$\int_{\Gamma_1 \cup \tilde{\Gamma}_2} f\,dx + g\,dy = \int_{\Gamma_1} + \int_{\tilde{\Gamma}_2} f\,dx + g\,dy$$
$$= \int_{\Gamma_1} f\,dx + g\,dy - \int_{\Gamma_2} f\,dx + g\,dy$$
$$= 0.$$

This shows that

$$\int_{\Gamma_1} F = \int_{\Gamma_2} F$$

for special pairs of curves Γ_1 and Γ_2. By modifying this proof to include all pairs of curves we can prove that F has a potential.

We now turn to complex analysis where Green's theorem and the Cauchy-Riemann equations can be combined to prove a very important theorem in complex analysis—Cauchy's theorem.

If $f = u + iv$ is a holomorphic function on an open set Ω in $\mathbf{C}$ and Γ is a parametrized curve in Ω we define

$$\int_\Gamma f(z)\,dz = \int_\Gamma (u+iv)(dx + i\,dy)$$
$$= \int_\Gamma u\,dx - v\,dy + i\int_\Gamma u\,dy + v\,dx.$$

Example 77. We evaluate $\oint_{|z|=r} \dfrac{dz}{z^n}$, where n is an integer. The mapping $P(\theta) = re^{i\theta}$, $\theta \in [0, 2\pi]$, is an anticlockwise parametrization of $|z| = r$. By de Moivre's theorem

$$P(\theta) = r(\cos\theta + i\sin\theta).$$

Hence $x = r\cos\theta$, $y = r\sin\theta$ and

$$\oint_{|z|=r} \frac{dz}{z^n} = \int_0^{2\pi} (re^{i\theta})^{-n}(-r\sin\theta + ir\cos\theta)\,d\theta$$
$$= \int_0^{2\pi} (re^{i\theta})^{-n} \cdot ire^{i\theta}\,d\theta$$
$$= i\int_0^{2\pi} r^{-n+1} e^{i(1-n)\theta}\,d\theta$$
$$= \begin{cases} 2\pi i & \text{if } n = 1 \\ \left[ir^{-n+1}\dfrac{e^{i(1-n)\theta}}{i(1-n)}\right]_0^{2\pi} = 0 & \text{if } n \neq 1. \end{cases}$$

Green's Theorem

In the course of working out this example, from the definition, we note that the term dz in the original integral is replaced by $ire^{i\theta}\,d\theta$ in the "parametrized" integral. This is not surprising since $z = re^{i\theta}$ implies $\dfrac{dz}{d\theta} = ire^{i\theta}$. Many such obvious simplifications are valid but one must always verify them mathematically.

We now prove Cauchy's theorem. In the proof we assume that the real and imaginary parts of a holomorphic function f have continuous partial derivatives. This is always true but we have not proved it. We assume that the curve Γ admits a nice parametrization and that the interior of Γ is a type I and type II domain—in order to apply our version of Green's theorem. This is quite a lot of assumptions. Many of these assumptions can be removed by further analysis. We note, however, that Example 77 shows that one of the hypotheses given in our statement of Cauchy's theorem cannot be removed. Identify it.

Cauchy's theorem. *If Γ is a closed curve and f is holomorphic on an open set U which contains Γ and all points "inside" Γ, then*

$$\oint_\Gamma f(z)\,dz = 0.$$

Proof. By Green's theorem

$$\oint_\Gamma f(z)\,dz = \oint_\Gamma u\,dx - v\,dy + i\int_\Gamma u\,dy + v\,dx$$

$$= \int\int_\Omega \left(-\frac{\partial v}{\partial x} - \frac{\partial u}{\partial y}\right) dx\,dy + i\int\int_\Omega \left(\frac{\partial u}{\partial x} - \frac{\partial v}{\partial y}\right) dx\,dy$$

where Ω is the inside of Γ. By the **Cauchy-Riemann equations** we have $\dfrac{\partial u}{\partial x} = \dfrac{\partial v}{\partial y}$. Hence $\dfrac{\partial u}{\partial x} - \dfrac{\partial v}{\partial y} = 0$ and the second integral is zero.

The second Cauchy-Riemann equation implies

$$\frac{\partial u}{\partial y} = -\frac{\partial v}{\partial x} \quad \text{and hence} \quad -\frac{\partial v}{\partial x} - \frac{\partial u}{\partial y} = 0.$$

Hence the first integral is also zero. This proves the theorem.

The converse to this theorem is also true and is known as **Morera's theorem:** if $f = u + iv$ and $\int_\Gamma f(z)\,dz = 0$ for every simple closed curve in Ω, then f is holomorphic on Ω.

A consequence of Cauchy's theorem is that line integrals of holomorphic functions only depend on the end points. Specifically, if Γ_1 and Γ_2 are two directed curves in Ω with the same initial point a and the same final point b (Figure 22.4) then

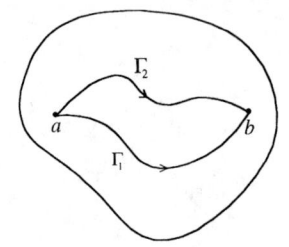

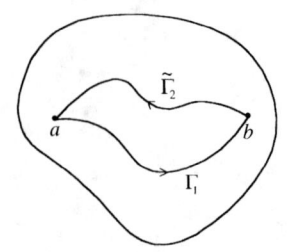

$$\int_{\Gamma_1} f(z)\,dz = \int_{\Gamma_2} f(z)\,dz.$$

Figure 22.4

From Γ_1 and Γ_2 we construct a closed curve Γ by first taking Γ_1 and adding to it the curve $\tilde{\Gamma}_2$ which is Γ_2 with its direction reversed (Figure 22.4). By Cauchy's theorem

$$\int_{\Gamma} f(z)\,dz = 0 = \int_{\Gamma_1} f(z)\,dz + \int_{\tilde{\Gamma}_2} f(z)\,dz$$
$$= \int_{\Gamma_1} f(z)\,dz - \int_{\Gamma_2} f(z)\,dz.$$

Hence

$$\int_{\Gamma_1} f(z)\,dz = \int_{\Gamma_2} f(z)\,dz.$$

It is of course possible that Γ does not admit a smooth parametrization, see Figure 22.4. To overcome this difficulty we should really introduce the idea of a "piecewise smooth" curve, i.e., a curve obtained by joining end to end a finite number of smooth non-closed curves. This would have been helpful on a number of occasions, e.g., in the proof of Theorem 67 and in the above example, and would have enabled us to prove more general versions of Green's theorem and Cauchy's theorem. Such a concept would also have allowed us to discuss curves with corners. The essential ideas and methods, however, would remain the same and we have not developed this degree of generality in order to avoid further technicalities that might tend to obscure the essentials.

It is also possible that Γ_1 and Γ_2 cross one another. In this case it is not possible to apply directly our version of Cauchy's theorem. A more general form of Cauchy's theorem is true and this can be applied to curves Γ_1 and Γ_2 which intersect one another.

Green's Theorem

To complete our discussion we use vector notation. We suppose that Γ is a closed curve and that the inside of Γ, Ω, is an open set of type I and type II. Let $P : [0, l] \to \Gamma$ denote a unit speed (anticlockwise) parametrization of Γ. If $P(s) = (x(s), y(s))$, then $T(s) = (x'(s), y'(t))$ and $N(s) = (-y'(s), x'(s))$. Since P is an anticlockwise parametrization and N is obtained by rotating T through an angle $+\pi/2$ it follows that N points into Ω. For this reason we call N the **inner normal**. The **outer normal**, which we denote by n, is equal to $-N$. We have

$$n(s) = N(s) - (y'(s), -x'(s)).$$

Now suppose $F = (f, g) : \mathbf{R}^2 \to \mathbf{R}^2$. Since $F(x,y)$ is a vector for each point (x, y), such a mapping is often called a **vector field**. The divergence of F, which is written as div (F) is defined as follows:

$$\text{div}(F) = \text{div}(f, g) = \frac{\partial f}{\partial x} + \frac{\partial g}{\partial y}.$$

We have

$$\oint_\Gamma (F \cdot n)\, ds = \int_0^l \Big(f(P(s)) \cdot y'(s) + g(P(s)) \cdot (-x'(s))\Big)\, ds$$

$$= \int_0^l \Big(-g(P(s)), f(P(s))\Big) \cdot P'(s)\, ds$$

$$= \int_\Gamma -g\, dx + f\, dy$$

$$= \int\int_\Omega \left(\frac{\partial f}{\partial x} + \frac{\partial g}{\partial y}\right) dx\, dy, \quad \text{by Green's theorem.}$$

This result is known as the divergence theorem and admits physical interpretations in heat conduction and fluid mechanics.

Divergence theorem. *If the closed curve Γ is the boundary of an open set Ω and F is a vector field with continuous first order partial derivatives on an open set containing Γ and Ω, then*

$$\oint_\Gamma (F \cdot n)\, ds = \int\int_\Omega (\text{div}\, F)\, dx\, dy.$$

If $\phi : \mathbf{R}^2 \to \mathbf{R}$ is differentiable, then $\nabla \phi$ is a vector field. If ϕ is harmonic, then

$$\text{div}(\nabla \phi) = \text{div}\left(\frac{\partial \phi}{\partial x}, \frac{\partial \phi}{\partial y}\right) = \frac{\partial^2 \phi}{\partial x^2} + \frac{\partial^2 \phi}{\partial y^2} = 0$$

and the divergence theorem shows that

$$\oint_\Gamma (\nabla \phi \cdot n)\, ds = 0.$$

Exercises

22.1 Parametrize the curve $x^{2/3} + y^{2/3} = a^{2/3}$ and use Green's theorem to calculate the area enclosed by this curve. Use Lagrange multipliers to find the radius of the largest circle centered at the origin which lies inside this curve.

22.2 Show that the area of the region bounded by the rays $\theta = \alpha$ and $\theta = \beta$ and the curve $r = f(\theta)$ equals
$$\frac{1}{2} \int_\alpha^\beta f(\theta)^2 \, d\theta.$$

22.3 Evaluate, for $a = 1$ and $a = 3$,
$$\oint_{|z-a|=2} \frac{dz}{z}.$$

22.4 Let f and g denote harmonic functions from $\mathbf{R}^2 \to \mathbf{R}$. Let Γ denote a closed curve and let n denote the outer normal to Γ. Use the divergence theorem to show
$$\oint_\Gamma f \frac{\partial g}{\partial n} \, ds = \oint_\Gamma g \frac{\partial f}{\partial n} \, ds.$$

22.5 If f and g are $\mathbf{C}$-differentiable functions defined on an open subset U of $\mathbf{R}^2$ and Γ is a closed curve in U which encloses the set Ω show that
$$\int_\Gamma \overline{f(z)} \, g(z) \, dz = 2i \int \int_\Omega \overline{f'(z)} \, g(z) \, dx \, dy.$$

22.6 Write down a list of five non-rigorous remarks from each of the last two chapters. By consulting the literature find a rigorous proof for one of them.

Solutions

Ch. 1
1.1 (a) $x \neq 0$ (b) $u \neq 2v$ (c) $\{s > 0, 0 < r < 1\} \cup \{s < 0, r < 0\}$.
1.2 (a) $[-1, +1]$ (b) $[-2, +2]$ (c) $[-2, +2]$. **1.4** Use exercise 1.10
1.6 -16 **1.9** (a), (b) open.

Ch. 2
2.1 (a) circles about the origin (b) straight lines (c) and (d) ellipses
2.2 see 3.2. **2.3** $c = 4$ **2.4** (a) 0, (b) -1, (c) -1, (d) -3.

Ch. 3
3.1 $P = (1, 2)$, $\vec{v} = (2, 3)$ (a) 189 (b) $5/4$ (c) $5e^2 + 4\cos 1 + 3\sin 1$
3.2 (a) $(8x^3y^3 - y^2, 6x^4y^2 - 2xy + 3)$
(b) $(\frac{1}{y} + \frac{y}{x^2}, -\frac{x}{y^2} - \frac{1}{x})$ (c) $(e^y + y\cos x, xe^y + \sin x)$
3.3 (a) $(0,0), (1,1)$ (b) $\pm((2+\sqrt{3})^{\frac{1}{2}}, (2-\sqrt{3})^{\frac{1}{2}})$, $\pm((2-\sqrt{3})^{\frac{1}{2}}, (2+\sqrt{3})^{\frac{1}{2}})$,
(c) $(0,0), (0, \frac{3}{4}), (\frac{9}{4}, 0), (\frac{9}{4}, \frac{3}{4})$ **3.4** 5, $(-4, 3)$, $(4, -3)$.

Ch. 4
4.2 (a) $(0,0)$ (b) $(\frac{1}{3}, \frac{2}{3})$, $(-1, -2)$ (c) No critical points (d) $(0,0)$, $(-\frac{1}{4}, -\frac{1}{2})$
(e) $\nabla f = (\sin y \sin(x + 2y), \sin x \sin(2x + y))$, $(n\pi, m\pi), \pm(\frac{\pi}{3} + n\pi, \frac{\pi}{3} + m\pi)$,
n, m integers (f) $(1, 1)$ **4.5** $\left(\frac{x+1}{5}\right)^2 + \left(\frac{y-2}{3}\right)^2 = 1$.

Ch. 5
5.1 (from 3.3) (a) saddle point at $(0,0)$ local minimum at $(1, 1)$, (b) saddle points at $\pm((2 - \sqrt{3})^{1/2}, (2 + \sqrt{3})^{1/2})$ local maximum at $(-(2 + \sqrt{3})^{1/2}, -(2 - \sqrt{3})^{1/2})$, local minimum at $((2 + \sqrt{3})^{1/2}, (2 - \sqrt{3})^{1/2})$, local minimum at $(9/4, 9/4)$, (from 4.2) (d) local minima at $(0, 0)$ and saddle point at $(-\frac{1}{4}, -\frac{1}{2})$, (e) saddle points at $\pm(\pi/3 + n\pi, \pi/3 + m\pi)$, n, m integers (f) local minimum at $(1, 1)$ **5.2** local maximum at $(\frac{\pi}{3}, \frac{\pi}{3})$ **5.3**

$a = \frac{33}{5}, b = \frac{123}{5}$ **5.4** 100, 300, 170000 **5.5** max 1, min -1 **5.6** minimum 1 occurs at $\{(x,0); 0 \leq x \leq 1\}$.

Ch. 6
6.2 (a) $-\frac{2}{3}$ (b) 0 (e) 0 (f) 0 (c) and (d) do not exist.

Ch. 7
7.1 100 **7.2** 2.9, 0.0672, 2.24% **7.4** 0.28, 0.017.

Ch. 8
8.2 (a) $9y + 8x + 25 = 0$, $9x - 8y + 10 = 0$ (b) $5x + 2y = 9$, $2x - 5y = 50$.
8.3 A triangle **8.4** $y > 0$, $y = 3(1-\frac{x^2}{4})^{1/2}$, $y < 0$, $y = -3(1-\frac{x^2}{4})^{1/2}$, $x > 0$, $x = 2(1-y^2/9)^{1/2}$, $x < 0$, $x = -2(1-y^2/9)^{1/2}$ **8.5** $g'(x) = \frac{1}{2}e^{2x} - \frac{1}{9}\sin 2x$.

Ch. 9
9.1 (a) $5, 0$ (b) minimum $\frac{7}{32}$, no maximum (c) $25, -25$, (d) $1, \frac{1}{2}$ **9.2** $5, -\frac{1}{3}$
9.4 $\sqrt{3}, -\sqrt{3}$ **9.5** $8, -1$ **9.6** (a) $(\sqrt{2}r, r/\sqrt{2})$ (b) $(\sqrt{2}r, \sqrt{2}r)$ (c) $(\sqrt{2}a, \sqrt{2}b)$
9.7 $4/\sqrt{5}$ **9.10** $3, 1$ **9.11** 0.

Ch. 10
10.1 (a) $0, 67$ (b) $-27, 45$ (c) $-1, 13$ **10.3** $T(-\frac{1}{7}, \pm\frac{8\sqrt{3}}{7}) = 8\frac{1}{7}$, $T(1,0) = -1$
10.4 $G_1 = 17$, $G_2 = 13$, marginal profit 45.

Ch. 11
11.1 (a) $16x + 18y + z + 25 = 0$, $(16t - 2, 18t - 1, t + 25)$
(b) $40x + 16y - z = 36$, $(5 + 40t, -8 + 16t, -t + 36)$
(c) $x + \sqrt{3}y + z = (\pi + \sqrt{3})/\sqrt{3}$, $(t, \frac{\pi}{3} + t\sqrt{3}, t + 1)$
(d) $x + \frac{1}{2}y - z = 2 - \log 2$, $(1 - 2t, 2 - t, 2t + \log 2)$ **11.2** $(\frac{7}{8}, -\frac{5}{18}, \frac{541}{144})$ **11.3** $x - 2y + z = 0$ **11.4** $3, \frac{5}{2}, 2$.

Ch. 12
12.1 $-1, -2^{7/3}$ **12.2** $e^t(\sin^2 t + \sin 2t) - 2e^{2t}$ **12.3** $f \circ g = (1, \frac{1}{2}\sin xy)$, $g \circ f = (\sin(x^3y + xy^3), \cos(x^3y + xy^3))$ **12.4** $\cos x f_x + (e^x - \sin x) f_y$ **12.5** 11 **12.8** -56×10^{-6}.

Ch. 13
13.1 $(2, 1)/\sqrt{5}$, $(-1, 2)/\sqrt{5}$ **13.2** $3(\sqrt{2} - 1)/4\sqrt{2}$ **13.3** $\sqrt{2}(e - 1)$ **13.5** (a) $(\frac{t}{\sqrt{2}}\cos(\log(\frac{t}{\sqrt{2}})), (\frac{t}{\sqrt{2}}\sin(\log(\frac{t}{\sqrt{2}}))$, $t \in [\sqrt{2}, \sqrt{2}e^{\frac{\pi}{2}}]$ (b) $(\cos t, \sin t)$, $t \in [0, \pi]$ (c) $(\sinh^{-1}(t), \sqrt{1+t^2})$, $t \in [0, \sinh 1]$.

Ch. 14
14.1 $e^x(1 + e^{2x})^{-3/2}$, $2|xy| \cdot |x^3 - y^3|(x^4 + y^4)^{-3/2}$ **14.2** $(4\cosh t, 9 \sinh t)$, $-36(4\sinh^2 t + 81\cosh^2 t)^{-3/2}$
14.3 $(9\cos t, 3\sin t)$, $27(81\sin^2 t + 9\cos^2 t)^{-\frac{3}{2}}$, $\pm(9, 0)$

14.6 $(2, \pm\sqrt{3})$, $(1,0)$, $(1,0)$ **14.7** $\left(\frac{9}{4}(x-2)^2 + \frac{4}{9}(y-3)^2\right)^{\frac{-3}{2}}/6$ **14.8** $\left(\frac{4x^2}{9c^2}\right) + \left(\frac{4y^2}{5c^2}\right) = 1$ **14.9** Maximum at $(0,0)$, minimum at $\left(\frac{1+\sqrt{17}}{8}, \frac{6\sqrt{17}-10}{64}\right)$.

Ch. 15
15.2 0.

Ch. 16
16.4 $(\pm\frac{1}{3\sqrt{2}}, 0), (0, \pm\frac{1}{5})$.

Ch. 17
17.2 $(e^x \cos y, e^x \sin y)$, $(\sinh x \cos y, \cosh x \sin y)$, $(\cosh x \cos y, \sinh x \sin y)$, $(e^x(x \cos y - y \sin y), e^x(y \cos y + x \sin y))$, $(x \cos x \cosh y + y \sin x \sinh y, y \cos x \cosh y - x \sin x \sinh y)$
17.5 $\frac{1}{2}(r + r^{-1})\cos\theta, \frac{1}{2}(r - r^{-1})\sin\theta$.

Ch. 18
18.1 $-\frac{1}{4}, \frac{3\pi}{32}$ **18.2** (a) 21, 14 (b) $\frac{1}{12}, \frac{5}{18}$ (c) $2(\cos\frac{1}{2} - \cos 1)$
18.3 $3(e-1)/4$, $e^{-1} - e^{-2}$ **18.4** 0.

Ch. 19
19.1 (a) $\frac{34}{15}$ (b) 0 **19.3** (a) $\sin(xy)$ (b) $x^2 + xy + e^{x^2} + e^{y^2}$ (c) does not exist (d) $x^2 y + xy^3 + \sin(x^2)$ **19.4** $b = \sqrt{\left(\frac{3a}{2}\right)}$ **19.5** $e^x(y\cos y + x \sin y), ze^z$.

Ch. 20
20.1 $(\pi - 2)/24$ **20.2** $-\frac{3\pi}{2}$ **20.3** $\frac{32}{3}$ **20.4** $(e^4 - 1)/4$, $(e-2)/2$.

Ch. 21
21.1 (a) $\pi/2$ (b) $\pi(b^2 - a^2)/2$ (c) $3\pi/2$ **21.3** $\frac{8}{45}$
21.5 $u(r,\theta) = re^{r\cos\theta}\cos(\theta + r\sin\theta)$, $v(r,\theta) = re^{r\cos\theta}\sin(\theta + r\sin\theta)$.

Ch. 22
22.1 $3a^2\pi/8$, $a/2\sqrt{2}$ **22.3** $2\pi i, 0$.

Index

absolute curvature 107, 117, 123
angle 107, 136, 165
anticlockwise 99–100, 116, 129, 175, 180, 183
appropriate domain 153, 179
approximation 46, 51, 57, 81, 90, 119, 171
area 50, 69, 71, 139, 157, 171, 177, 184
average 18, 51, 139

bijective function 103, 172
binomial theorem 134
boundary 2–4, 44, 61–62, 65, 69, 80, 98, 148, 159, 178
bounded set 43, 65

C-differentiable function 133, 174, 184
capital 52, 77
Cartesian coordinates 165, 169
Cauchy-Riemann equations 134–135, 168, 174, 180–181
Cauchy-Schwarz inequality 53
Cauchy's theorem 181
centre of curvature 123, 128
chain rule 58, 75, 87, 91, 104, 108, 115, 120, 149
change of variable 96, 124, 166, 168, 172
circle of curvature 122–123
closed curve 99, 116, 130, 175, 183
    set 43–44, 65, 105, 181, 183
Cobb-Douglas function 77
column vector 24, 90
completing the square 25, 117

complex numbers 132
    temperature 137
    variable 133
composition of functions 91, 103
cone 111
conic sections 111
connected open set 149
conservative force 154
constraint 62, 76, 78
continuous function 1, 10, 23, 38, 43–44, 103, 105, 111, 139, 148, 157–158, 167, 172
contour 9
convex curve 130–131
convex set 131
coordinate system 165, 168
coordinate mappings 166
corner 98, 101, 130, 181
cost 36, 80
counterclockwise 99
critical points 2, 16, 20–21, 28, 30–31, 34–35, 43, 54, 74, 120, 131
cross section 8–9, 17, 82–83, 111
current 97
curvature 107, 123, 127
curve 9, 38, 42, 55, 64, 89, 99, 121, 142, 149–150, 154, 179, 182–183

degenerate critical points 30
demand (price) 80
de Moivre's theorem 180
derivative 4, 12, 90
determinant (= det) 27, 30, 79, 171–172

Index 189

differentiable function 46, 56, 90, 103, 118, 125, 133, 148
differential geometry 130
directed curve 99, 116, 142–143, 145, 149, 166
direction 11, 14, 38, 54–55, 99, 113, 116, 182
directional derivative 17, 20, 51–52, 89
disc 2, 16, 94, 121, 172
distance 66, 102, 117, 123, 144, 164
divergence theorem 183
domain 96, 139, 153, 157, 161
domain of integration 148
dot product 18, 83–84, 86, 126–127, 135, 144
double integral 156, 171

eigenvalue 79
eigenvector 79
electrostatics 137
ellipse 66, 70, 109, 111, 115, 117, 130, 137, 145, 151, 154, 177
error 46, 50–51, 59, 120
expected value 139

final point 99, 109, 113, 145, 147, 182
finance 77
first order partial derivative 37, 96, 149, 151, 172, 173, 183
fluid mechanics 183
flux 137
force field 154
four vertex theorem 130
Fubini's theorem 160
function
    bijective 103, 172
    C-differentiable 133, 174, 184
    Cobb-Douglas function 77
    continuous 1, 10, 23, 38, 43–44, 103, 105, 111, 139, 148, 156–157, 167, 172
    differentiable 46, 56, 90, 103, 118, 125, 133, 148
    harmonic 28, 135, 155, 168–169, 174, 183–184
    holomorphic 133, 174, 180–181
    hyperbolic 111
    injective 99, 166, 174

integrable 138, 157, 169, 172
inverse 103, 163
moment generating 164
monotonic 167
objective 62
Riemann integrable 138
stream 137
sufficiently regular 9–10, 16, 21–24, 31, 37, 62
symmetric 34
trigonometric 111
fundamental theorem of calculus 103, 119, 148–149, 176
    existence theorem 43, 78

generalized sequence 139
geometric curve 99
gradient 16, 67, 83, 89, 148
graph 5, 8, 31, 54–56, 59, 73, 74, 81, 85, 98, 113–114, 116, 121, 139, 145, 147, 157, 160, 163, 176
Green's theorem 148, 175, 179, 183–184

harmonic function 28, 135, 155, 168–169, 174, 183–184
    conjugate 137, 154
heat 137, 183
Hessian 24, 27, 35, 75, 90
holomorphic function 133, 174, 180–181
hyperbola 111, 136
hyperbolic function 111

imaginary part 132–133, 135, 174
implicit differentiation 95
    function theorem 56, 73–74, 95, 114
initial point 99, 109, 113, 145, 147, 182
injective function 99, 166, 174
inner product 18
    normal 183
integral 138, 156, 178
integrable function 138, 157, 169, 172
intermediate value theorem 111
inverse function 103, 163
isotherm 137

Jacobian 171

labour 77

Lagrange multipliers 4, 27, 32, 43, 65, 72, 74–75, 77–78, 94, 96, 113, 184
Laplace's equation 135
least squares estimate 35
   regression line 35–36
Lebesgue integral 139
length 19–20, 50, 69, 101–102, 167, 171
level curve 9
   set 9–10, 55, 59, 62, 64, 68, 74, 98, 112, 114, 124, 136, 165
l'Hôpital's rule 40
limit 4, 38, 40–41, 46–47, 122, 138, 141, 157–158, 171
limits of integration 142, 158
line 22, 35, 41–42, 47, 129
   integral 138, 145, 148, 167, 179
   normal 60, 84, 101
   tangent 5, 47, 57, 60, 64, 81, 84, 101, 130
linear algebra 25, 78
   approximation 47, 51, 57, 81, 118
   model 35
   programming 71
local maximum 16, 22, 27, 31, 64, 73, 120, 130
   minimum 16, 23, 27, 31, 73, 120, 130

marginal productivity 78
   profit 36, 52
matrix 24, 79, 86, 90
   multiplication 24, 90–91
   notation 24, 91
maximum 1–2, 10–12, 14–16, 30–31, 43, 55, 72, 105, 131
   curvature 110, 130
   increase 67
mean value theorem 48, 52, 119
minimum 1–2, 16, 31, 43, 72, 105
minimum curvature 110, 130
mixed partial derivative 23, 37, 97
model 35, 136
moment generating function 164
money 78
monopoly 80
monotonic function 167
Morera's theorem 181
mountain pass 31

negative curvature 129
net 139
non-degenerate critical point 30
normal 68, 100, 104, 123, 128, 136
   inner 183
   outer 183
   unit 100, 127
normal line 60, 84, 101
notation 15, 24, 47, 88, 91

objective function 62
Ohm's law 97
open set 2, 15, 23, 31, 43, 52, 55, 61, 62, 64, 97, 132, 149, 156, 172, 175, 183
orthogonal curves 136
outer normal 183
oval 130

parabola 96, 111, 154
paraboloid 86
parallel vectors 64, 69, 74
parallelogram 170
parametrization 99, 102, 104, 114, 142–143, 150, 167, 180
parametrized curve 99, 105, 129, 149
partial derivative 12, 15, 17, 23, 27, 39, 45, 75, 89, 135, 149, 172, 175, 183
partition 138, 142, 156–157, 169
path 11, 69, 151
perimeter 69
perpendicular vectors 19, 60, 69, 74, 84, 86, 100, 124, 127, 136
piecewise differentiable curve 98, 182
plane 8, 81, 85
   curve 130
polar coordinates 166, 168, 172–173
position 99
positive curvature 129
potential 149, 152–154, 179–180
price (demand) 80
product rule 125
profit 36, 52, 80
Pythagoras' theorem 19

quadratic approximation 119

random variable 139
range 95, 139

Index 191

rate of change 58–59, 96
real part 132–133, 135, 169, 174
rectangle 36, 70, 156, 161, 170
regularity conditions 1, 10, 18, 37, 171
resistance 53, 97
Riemann sum 138, 140, 156–158, 170
    integrable 138
row vector 24, 90

saddle points 31, 75
scalar product 18
second derivative 6–7, 90, 118
second order directional derivative 21, 23, 37
    partial derivative 22, 27, 37, 45, 97, 120, 131, 155
sequence 38, 44, 139
silent variable 92
sink 136
slope 5, 16, 19, 60
source 136
speed 88, 101–103, 144
steady state 136
Stokes' theorem 175
stream function 137
    line 137
sufficiently regular function 9–10, 16, 21–24, 31, 37, 62
surface 8
symmetric function 34

tangent 100, 121, 136, 144, 154
    line 5, 47, 57, 60, 64, 81, 85, 100, 128, 137, 144
    plane 81, 85
    space 85
temperature 80, 94, 97, 136
thermal energy 136
theorem
    binomial 134

    Cauchy's 181
    de Moivre's 180
    divergence 183
    four vertex 130
    Fubini's 160
    fundamental theorem of calculus 103, 119, 148–149, 176
    fundamental existence 43, 78
    Green's 148, 175, 179, 183–184
    implicit function 56, 73–74, 95, 114
    intermediate value 111
    mean value 48, 52, 119
    Morera's 181
    Pythagoras' 19
    Stokes' 175
time 89, 99, 102, 144
trace 79
transpose 25, 86, 90
triangle 69, 147, 164
trigonometric function 111
type I domain 157, 176, 181
    II domain 159, 176, 181

uniform continuity 139
unique (linear) approximation 47
unit
    normal 100, 127
    speed parametrization 102, 107–108, 126–127, 168, 183
    tangent 100, 127, 145, 154
    vector 18, 100

vector field 183
velocity 101
vertex 130
voltage 97
volume 157

weighted average 18
work 154

$x \geq y$

$x - y \geq y$

$y = -40$
$x = 20$

$x - (-40) \geq 0$

$-20 - (-40) \geq 0$